# ARRL's Portable Antenna Classics

## Thirty antennas to get you on the air from anywhere!

**Editor: Steve Ford, WB8IMY**

**Production Staff:** **Michelle Bloom, WB1ENT,** Production Supervisor, Layout
**Jodi Morin, KA1JPA,** Assistant Production Supervisor, Layout
**Sue Fagan, KB1OKW,** Graphic Design Supervisor, Cover Design
**David Pingree, N1NAS,** Senior Technical Illustrator

Background photo: Mike Corey, KI1U
Top left photo: RJS Photography
Top right photo: Jerry Clement, VE6AB

Printed in the USA



ISBN: 978-1-62595-034-5

First Edition
Fourth Printing

# Table of Contents

## HF Antennas

*80 METERS*


**1** The NJQRP Squirt by Joe Everhart, N2CX


*80 – 10 METERS*


**5** Zip Cord Antennas and Feed Lines for Portable Applications by William Parmley, KR8L

**8** The Octopus – A Four Band Antenna for Portable Use by Geoff Haines, N1GY

**11** 8-Band Backpacker Special by Jim Andera, WB0KRX


*80 – 2 METERS*


**14** Build a Portable Antenna by Robert Johns, W3JIP

**16** A Packable Antenna for 80 through 2 Meters by Dennis Kennedy, N8GGI


*60 – 10 METERS*


**18** The Ultimate Portable HF Vertical Antenna by Phil Salas, AD5X


*40 – 20 METERS*


**25** A Small, Portable Dipole for Field Use by Ron Herring, W7HD


*40 – 10 METERS*


**27** A Simple and Portable HF Vertical Travel Antenna by Phil Salas, AD5X

**31** The Miracle Whip: A Multiband QRP Antenna by Robert Victor, VA2ERY

**35** Build an HF Walking Stick Antenna by Robert Capon, WA3ULH


*40 – 6 METERS*


**38** A Ground-Coupled Portable Antenna by Robert Johns, W3JIP

**43** An Off Center End Fed Dipole for Portable Operation on 40 to 6 Meters by Kazimierz Siwiak, KE4PT


*20 METERS*


**46** A Portable Twin-Lead 20-Meter Dipole by Rich Wadsworth, KF6QKI


*20 AND 17 METERS*


**48** A Paint Pole Antenna by Anthony Salvate, N1TKS

*20 – 10 METERS*

50 Fishing for DX with a Five Band Portable Antenna by Barry Strickland, AB4QL
56 Super Duper Five Band Portable Antenna by Clarke Cooper, K8BP
59 A Portable Inverted V Antenna by Joseph Littlepage, WE5Y
63 A Portable 2-Element Triband Yagi by Markus Hansen, VE7CA
66 A Simple HF-Portable Antenna by Phil Salas, AD5X
68 The Shooter—A 3-Band Portable Antenna by E.W. Ljongquist, W4DWK/W1CQS
71 A Traveling Ham's Trap Vertical by Doug DeMaw, W1FB

*15 METERS*

75 The Black Widow – A Portable 15 Meter Beam by Allen Baker, KG4JJH

*10 METERS*

80 Two on 10 by Al Alvareztorres, AA1DO

## VHF/UHF Antennas

*6 METERS*

83 A Portable Two-Element 6 Meter Quad Antenna by Peter Rimmel, N8PR

*2 METERS*

86 The Take Down Yagi by Jerry Clement, VE6AB
90 Build a Portable Groundplane Antenna by Zack Lau, KH6CP/1
92 A Traveler's 2-Meter GP Antenna by Doug DeMaw, W1FB
96 A Portable Quad for 2 Meters by R.J. Decesari, WA9GDZ/6

*2 METERS AND 70 CENTIMENTERS*

99 The DBJ-2: A Portable VHF-UHF Roll-Up J-Pole Antenna for Public Service by Edison Fong, WB6IQN

## Masts and Other Supports

102 A Tilt-Over Antenna Mast Born of Necessity by Bruce Belling, N1BCB
104 A One Person, Safe, Portable and Easy to Erect Antenna Mast by Bob Dixon, W8ERD
107 The Fiberpole – Evolution of an Antenna Mast by Dave Reynolds, KE7QF
109 A Vehicle Mounted Mast for Field Operating by Geoff Haines, N1GY
111 A Portable Antenna Mast and Support for Your RV by Paul Voorhees, W7PV
113 Amateur Use of Telescoping Masts by R.P. Haviland, W4MB
117 Portable Antennas and Supports by Gary Sutcliffe, W9XT

## Appendix

119

# Foreword

Portable operating has become a hot-button issue in Amateur Radio today. Much of this interest is being driven by the fact that so many amateurs are now impacted by severe antenna restrictions imposed by communities and homeowner associations. When they can't permanently install antennas outside their dwellings, their only alternatives may be to either use temporary portable antennas that they can set up and take down easily, or operate in locations away from their homes.

There is also a growing segment of the amateur population that is passionate about the outdoors, be it fishing, hiking, camping and more. Many hams, young and old alike, are taking their radios into the field and they need compact, lightweight antennas for the journey.

*Portable Antenna Classics* is a collection of helpful articles harvested from more than three decades of *QST* magazine. In selecting the articles for this anthology, the emphasis was placed on antennas that offer acceptable performance while still being attractive in terms of size and weight. We have also included several articles on the topic of masts and supports for portable antennas.

I hope you'll find this book enjoyable and useful!

73 . . .
Steve Ford, WB8IMY
*QST* Editor / ARRL Publications Manager

By Joe Everhart, N2CX

# The NJQRP Squirt

This reduced-size 80-meter antenna is designed for small building lots and portable use. It's a fine companion for the Warbler PSK31 transceiver.

At one time, 80 meters was one of the more highly populated amateur bands. Lately, it has become significantly less popular because much DXing has moved to the higher frequencies and many suburban lot sizes are too small to accommodate a full 130-foot, λ/2 antenna for the band. That's unfortunate, because 80 meters has lots of potential as a local-communication band—even at QRP levels. The recently published Warbler PSK31 transceiver can serve as a great facilitator for close-in QRP communication without much effort.[1] What's really needed to complement the Warbler for this purpose is an effective antenna that fits on a small suburban plot. Because PSK31 (which the Warbler uses) is reasonably effective even with weak signals, we can trade off some antenna efficiency for practicality.

## What's a Ham to Do?

I investigated a number of antenna possibilities to come up with a practical solution. One intriguing candidate is the magnetic loop. Plenty of design information for this antenna is presented in *The ARRL Antenna Book* and at a number of Web sites.[2, 3] To obtain high efficiency, however, the loop must be 10 feet or more in diameter and built from ½-inch or larger-diameter copper pipe. The loop needs a very low-loss tuning capacitor and a means of carefully tuning it because of its inherently narrow bandwidth. Another configuration, the DCTL, may be a solution, but it's likely not very efficient.[4]

An old standby antenna I considered is the random-length wire worked against ground. If it is at least λ/4 long (a Marconi antenna) or longer, it can be reasonably efficient. Shorter lengths are likely to be several S units down in performance and almost any length end-fed wire needs a significant ground system to be effective. Of course, you may not need much of a ground with a λ/2 end-fed wire, but it's as long as a center-fed dipole.

Vertical antennas don't occupy much ground space, but suffer the same low efficiency as the end-fed wire if they are practical in size.

Probably the easiest antenna to use with good, predictable performance is the horizontal center-fed dipole. Unfortunately, as mentioned earlier, the usual 80-meter λ/2 dipole is too large for many lots. But all is not lost! The dipole can be reduced to about a quarter wavelength without much sacrifice in operation (see the sidebar, "Trade-Offs"). Furthermore, if the dipole's center is elevated and the ends lowered—resulting in an inverted **V**—it takes up even less room. This article describes just such a dipole: the NJQRP Squirt.

## V for Victory

You can think of the Squirt as a 40-meter, λ/2 inverted-**V** dipole being used on 80 meters. Figure 1 is an overall sketch of the antenna; Figure 2 is a photograph of a completed Squirt prior to erection. The Squirt has two legs about 34 feet long separated by 90° with a feed line running from the center. When installed, the center of the Squirt should be at least 20 feet high, with the dipole ends tied off no lower than seven feet above ground. This low antenna height emphasizes high-angle NVIS (Near Vertical-Incidence Skyware) propagation that's ideal for 80-meter contacts ranging from next door out to 150 or 200 miles. And

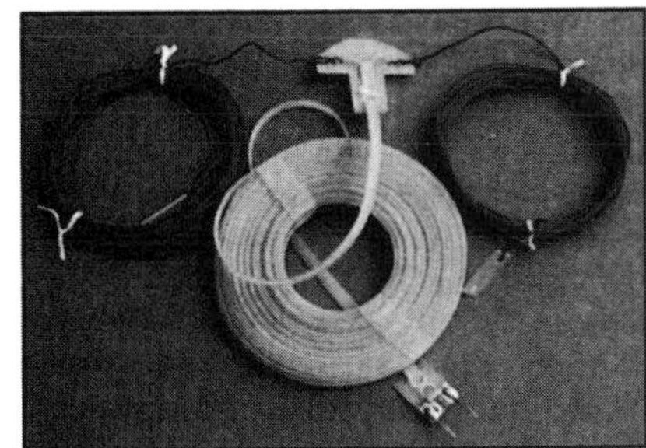

Figure 2—An assembled Squirt ready for installation.

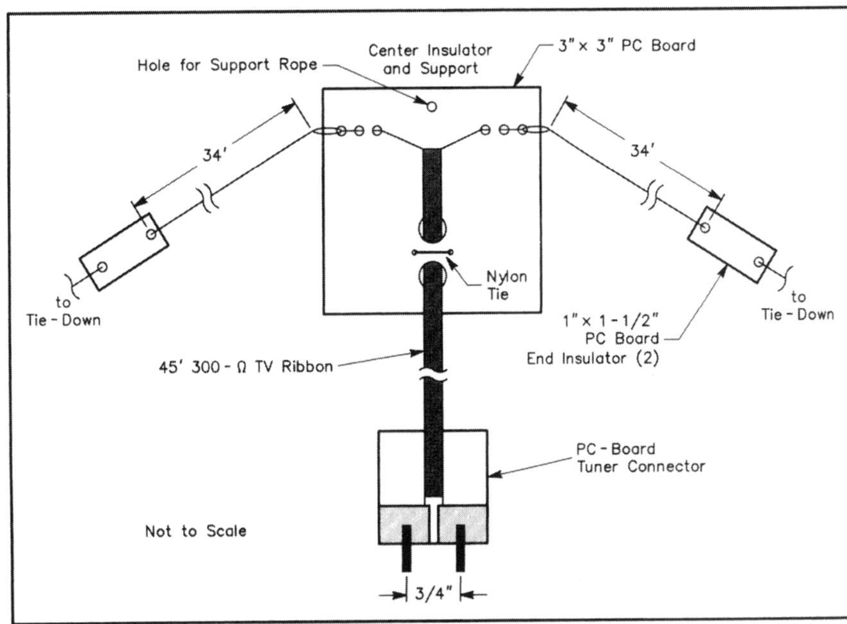


Figure 1—General construction of the 80-meter Squirt antenna.

[1]Notes appear on page 4.

that's where 80 meters shines! With the Squirt's center at 30 feet and its ends at seven feet, the antenna's ground footprint is only about 50 feet wide.

One nice feature of a λ/2 center-fed dipole is that its center impedance is a good match for 50- or 75-Ω coax cable (and purists usually use a balun). Ah! But the Squirt is only λ/4 long on 80 meters, so it *isn't* resonant! Its feedpoint impedance is resistively low and reactively high. This means that feeding the antenna with coax cable would create a high SWR causing significant feed-line loss. To circumvent this, we can feed the antenna with a low-loss feed line and use an antenna tuner in the shack to match the antenna system to common 50-Ω coax cable. I'll have more to say about the tuner later.

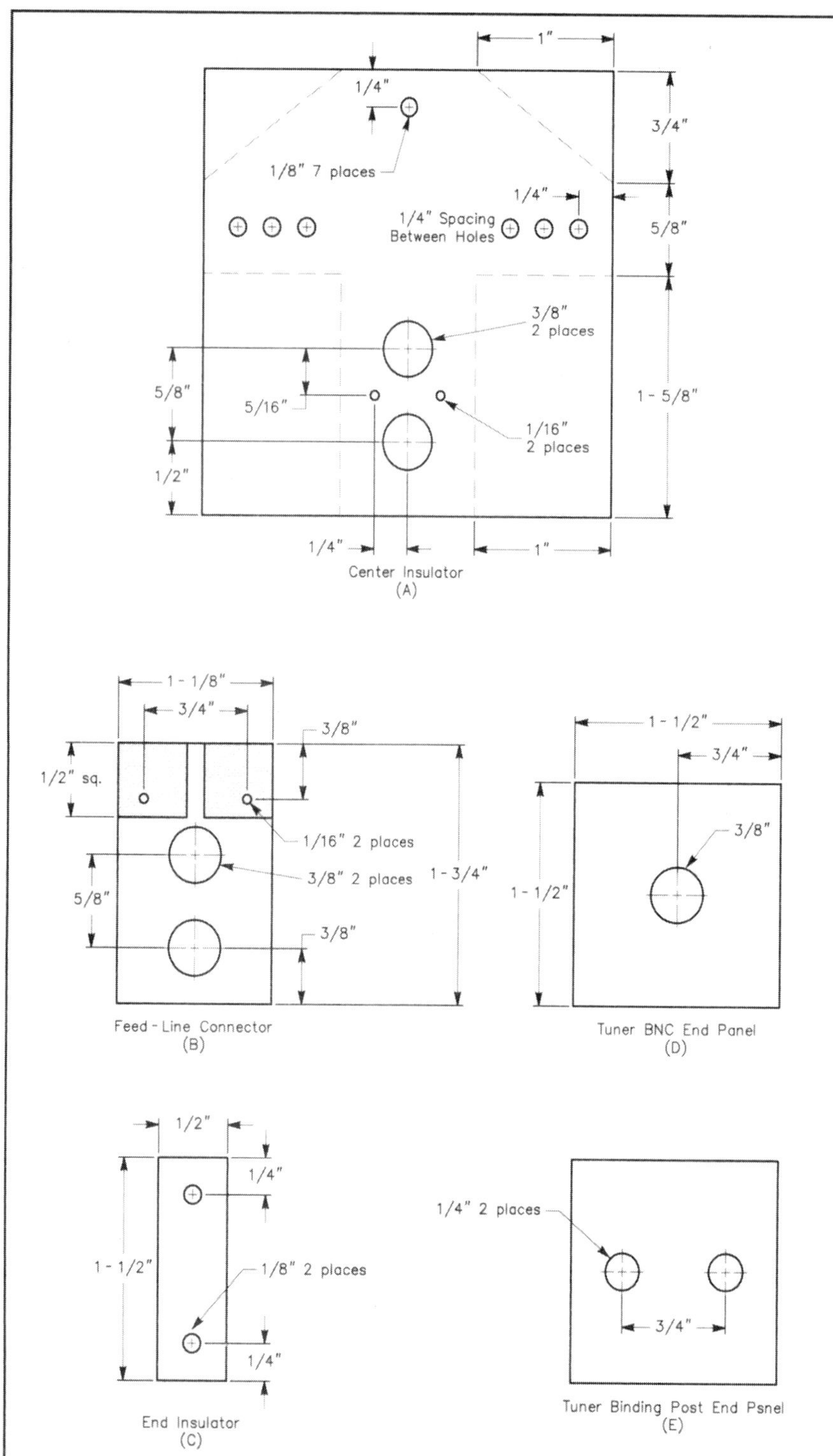


**Figure 3—Hole sizes and locations for the various PC-board pieces. See Note 5.**

I use 300-Ω TV flat ribbon line for the feed line. Although a better low-loss solution is to use open-wire line, that stuff is not as easy to bring into the house as is TV ribbon. Using TV ribbon sacrifices a little transmitted signal for increased convenience and availability. If you feel better using open-wire line, go for it!

## Using Available Materials

It's always fun to see what you can do with junkbox stuff, and this antenna is one place to do it. See the "Parts List" for information on materials and sources.[5] For instance, the end and center insulators (see Figure 1) are made of 1/16-inch-thick scraps of glass-epoxy PC board. For the antenna elements, I use #20 or #22 insulated hookup wire. Although this wire size isn't recommended for use with fixed antennas, I find it entirely adequate for my Squirt. Because it's installed as an inverted **V** antenna, the center insulator supports most of the antenna's weight making the light-gauge wire all that's needed. The small-diameter wire has survived quite well for several years at N2CX. This is not to say, of course, that something stronger like #14 or #12 electrical house wire couldn't serve as well.

The 300-Ω TV ribbon can be purchased at many outlets including RadioShack and local hardware stores. Once again, if you want to use heavier-duty feed line, do so. The only proviso is that you may then have to trim the feeder length to be within tuning range of the Squirt's antenna tuner.

### *The End Insulators*

I used 1/2×1 1/2-inch pieces of 1/16-inch PC board for the Squirt's two end insulators. As with everything else with the Squirt, these dimensions are not sacred; tailor them as you wish. If you use PC board for the end insulators, you have to remove the copper foil. This is easy to do once you've gotten the knack. Practice on some scraps before tackling the final product. The easiest way to remove the foil without etching it is to peel it off using a sharp hobby knife and needle-nose pliers. Carefully lift an edge of the foil at a corner of the board, grasp the foil with the pliers and slowly peel it off. You should become an expert at this in 10 or 15 minutes. Drill 1/8-inch-diameter holes at each end of each insulator for the element wires and tie-downs.

### *Tuner Feed-Line Connector*

The tuner end of the feed line is terminated in a special connector. Because the TV-ribbon conductors aren't strong, they'll eventually suffer wear and tear.

## Trade-Offs

One of the unfortunate consequences of shrinking an antenna's size is that its electrical efficiency is reduced as well. A full-size dipole is resonant with a feedpoint impedance that matches common low-impedance coax quite well. This means that most transmitter power reaches the antenna minus only 1 dB or so feed-line loss. However, when the antenna is shortened, it is no longer resonant. A *NEC-4* model for the Squirt shows that its center impedance on 80 meters is only about 10 Ω resistive, but also about 1 kΩ capacitive. This is a horrendous mismatch to 50-Ω cable, and feed-line loss increases dramatically with high SWR. The Squirt uses 300-Ω TV ribbon for the feed line with an inherently lower loss than coax. This loss is much less than if coax were used, but it's still appreciable. Calculated loss with 300-Ω transmitting feed line is about 7.7 dB (loss figures are hard to come up with for receiving TV ribbon) so the feed line used doubtless has more than that.

Although this sounds discouraging, it's *not fatal.* You have to balance losing an S unit or so of signal against not operating at all! Consider that the Squirt, even with its reduced efficiency, is still better than most mobile antennas on 40 and 80 meters. So for local communication (a low-dipole's forte), using PSK31 and the Squirt is quite practical.

If you don't already have an antenna, the Squirt's a good choice to get your feet wet when using PSK 31. Once you get hooked, you'll probably want a better antenna. If you have the room, put up a full-size dipole; you'll see the improvement right away. If you can't do that, use a lower-loss feeder on the Squirt, such as good-quality open wire.—*Joe Everhart, N2CX*

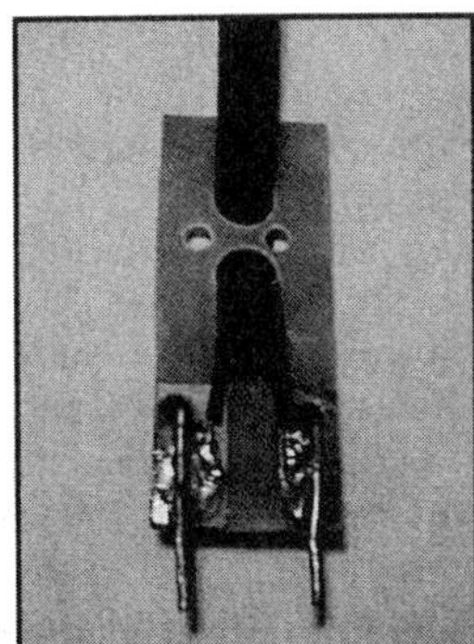

**Figure 4—The pad side of the home-made feed-line-to-tuner connector.**

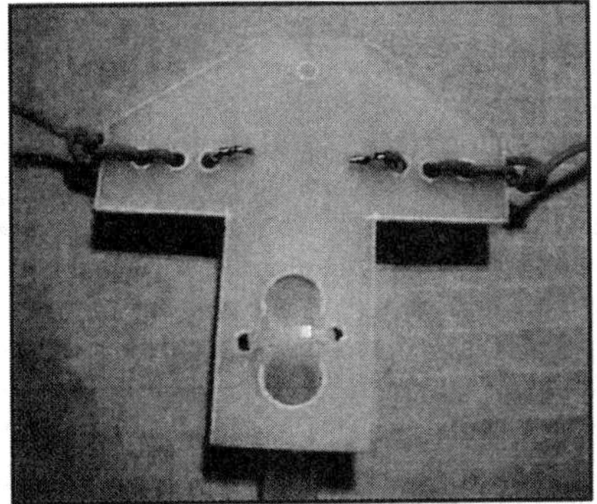

**Figure 5—Here the feed-line-to-tuner connector is shown attached to the binding posts of the Squirt antenna tuner.**

This connector provides needed mechanical strength and a means of easily attaching the feed line to the tuner. In addition to some PC-board material, you'll need four or five inches of #18 to #12 solid, bare wire. Refer to Figure 3 and the accompanying photographs in Figures 4 and 5 for the following steps.

Take a 1 1/8×1 3/4-inch piece of single-sided PC board and score the foil about 1/2 inch from one end; remove the 1 1/4-inch piece of foil. Now score the remaining foil so you can remove a 1/8-inch-wide strip at the center of the board, leaving two rectangular pads as shown in Figures 3B and 4. Drill two 1/16-inch-diameter holes in the copper pads spacing the holes about 3/4-inch apart. Drill two 3/8-inch holes at the connector midline about 5/8-inch apart, center to center, to pass the feed line and secure it.

Cut two pieces of #18 to #12 wire each about three inches long. Pass one wire through one of the 1/16-inch holes in the connector board and bend over about 1/4-inch of wire on the nonfoil side. Solder the wire to the pad on the opposite side and cut the wire so that about one inch of it extends beyond the connector. Repeat this procedure with the second wire. Next, strip about two inches of webbing from between the feed-line conductors and loop the feed line through the two 3/8-inch holes so that the free ends of the two conductors are on the copper-pad side. Strip each lead and solder each one to a pad. You now have a solid TV-ribbon connector that mates with the binding-post connections found on many antenna tuners. Figure 6 shows the connector mated with a Squirt tuner.

### *Center Insulator*

Strip all the foil from this 3-inch-square piece of board. Use Figure 3A as a guide for the hole locations. The top support hole and the six wire-element holes are 1/8-inch in diameter; space the wire-element holes 1/4-inch apart. The feed-line-attachment holes are 3/8-inch diameter spaced 1/2-inch apart, center to center; the two holes alongside the feed-line-attachment holes are 1/16-inch diameter. These 1/16-inch holes accept a plastic tie to secure the feed line. I trimmed the insulator shown in Figure 2 from its original 3-inch-square shape to be more esthetic. Your artistic sense may dictate a different pattern.

Bevel all hole edges to minimize wire and feeder-insulation abrasion by the glass-epoxy material. You can do this by running a knife around each hole to remove any sharp edges.

### Putting It All Together

The Squirt is simple to assemble. Once all the pieces have been fabricated, it should take no more than an hour or two to complete assembly. Begin with the center insulator. Cut each of the two element wires to a length of about 34 feet. Feed the end of one wire through the center insulator's outer hole on one side, then loop it back and twist around itself outside the insulator to secure it. Now loop it through the other two holes so that the inner end won't move from normal movement of the wire outside the insulator. Repeat the process for the other insulator/wire attachment. Separate several inches of the TV-ribbon feed-line conductors from the webbing; leave the insulation intact except for stripping about 1/2 inch from the end of each wire. Pass the TV ribbon through both 3/8-inch holes. Strip a 1/2-inch length of insulation from each dipole element, then twist each feeder wire and element lead together and solder the joints. It might be prudent also to protect the joint with some non-contaminating RTV or other sealant. Finally, loop a nylon tie through the holes alongside the feeder and tighten the tie to hold the feeder securely. A close-up of the assembled center insulator is shown in Figure 6.

Attach the end insulators to the free ends of the dipole wires by passing the wires through the insulator holes and twisting the wire ends several times to secure them.

So that the antenna/feed-line system can be tuned with the Squirt tuner, the 300-Ω feed line needs to be about 45 feet long. If you use a different tuner, you may have to make the feed line longer or shorter to be within that tuner's impedance-adjustment range.

### Tuner Assembly

This tuner (see Figures 7 and 8) is about as simple as you can get. It's a basic series-tuned resonant circuit link-coupled to a coaxial feed line. At C1, I use a 20 to 200-pF mica compression trimmer acquired at a hamfest (you *do* buy parts at hamfests, don't you?), although almost any small variable capaci-

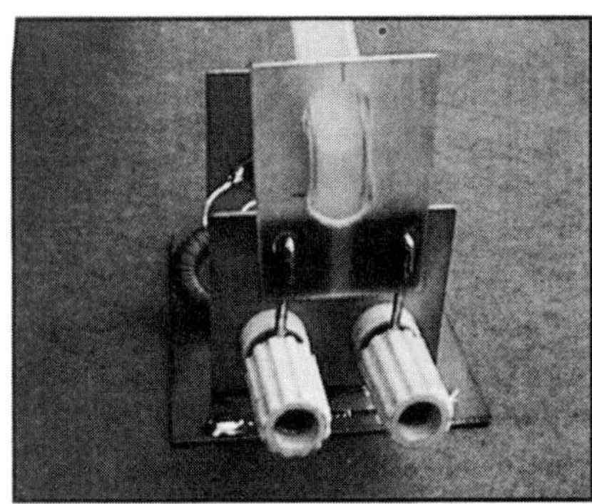

Figure 6—View of an assembled center insulator fashioned from a 3×3-inch piece of PC board from which all the foil has been removed.

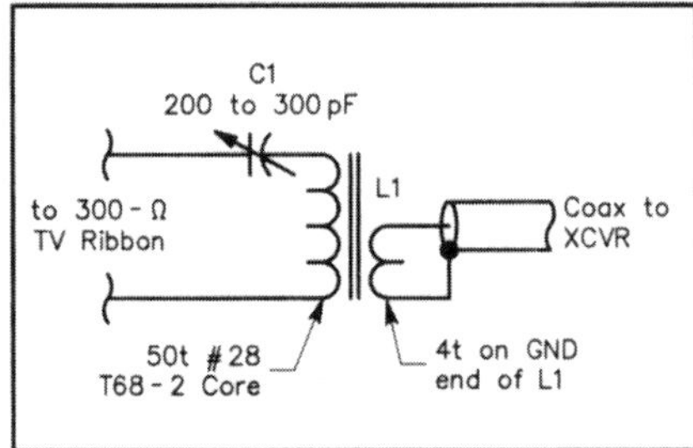


Figure 7—Schematic of the Squirt antenna tuner. See the accompanying Parts List.

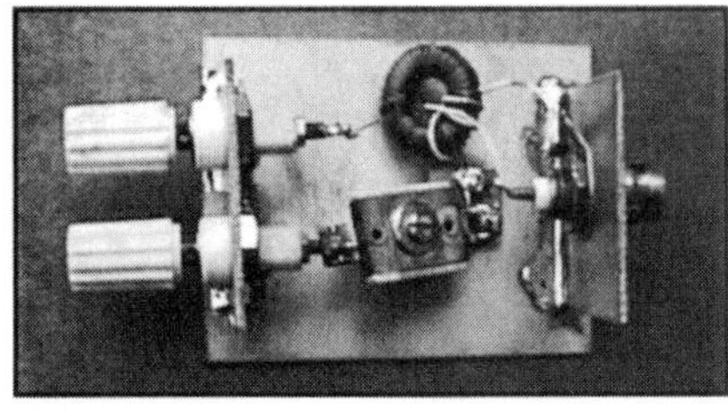

Figure 8—This Squirt tuner prototype uses a 2×3-inch piece of PC board for the base plate, two 1 1/2 × 1 1/2-inch pieces for end plates and a 1/2-inch square piece as a tie point for the toroid and tuning capacitor.

## Parts List

### Squirt Antenna

**Numbers in parentheses refer to vendors presented at the end of the list.**

1—3x3-inch piece of 1/16-inch-thick glass-epoxy PC board for the center insulator (1)
2—1/2x1 1/2-inch pieces of PC board for the end insulators (1)
1—1 1/8x1 3/4-inch piece of PC board for the feed-line connector (1)
2—34-foot lengths of #20 (or larger) insulated hookup wire (2)
1—6-inch length of #16 (or larger bare) copper wire; scrounge scraps from your local electrician.
1—45-foot length of 300-Ω TV ribbon line (2)

### Squirt Tuner

1—2x3-inch piece of PC board for base plate (1)
2—1 1/2-inch-square pieces of PC board for end plates (1)
1—1/2-inch-square piece of PC board for the tie point (1)
1—200- to 300-pF (maximum) mica compression trimmer (3)
1—T68-2 toroid core (3)
2—Five-way binding posts (2)
1—55-inch length of #26 or 28 enameled wire (2 and 3)

**Note**: You can use 3/16-inch-thick clear Plexiglas for the Squirt's end and center insulators. Commonly used as a replacement for window glass, Plexiglas scraps can be obtained at low cost from hardware stores that repair windows.

### Vendors

1. HSC Electronic Supply, 3500 Ryder St, Santa Clara, CA 95051; tel 408-732-1573, **www.halted.com**
2. Local RadioShack outlets or **www.RadioShack.com**
3. Dan's Small Parts and Kits, Box 3634, Missoula, MT 59806-3634; tel 406-258-2782; **www.fix.net/~jparker/dans.html**

tor of this value should serve. The inductor, L1, consists of 50 turns of enameled wire wound on a T68-2 iron-core toroidal form. An air-wound coil would do as well, although it would be physically much larger. Figure 8 shows the tuner built on an open chassis made of PC board. My prototype uses several PC-board scraps: a 2×3-inch piece for the base plate, two 1 1/2×1 1/2-inch pieces for each end plate (refer to Figure 3). A 1/2-inch square piece of PC board (visible just beneath the capacitor in Figure 8) is glued to the base plate to serve as an insulated tie point for the connection between the toroid (L1) and tuning capacitor (C1). The tuner end plates are soldered to the base plate to hold a pair of five-way binding posts and a BNC connector at opposite ends. L1 and C1 float above electrical ground, connected to the TV ribbon. One end of L1's secondary (or link) is grounded at the base plate and the coax-cable shield. The hot end of L1's secondary winding is soldered to the coax-connector's center conductor.

### Tuner Testing

C1 tunes sharply, so it's a good idea to check just how it tunes before you attach the tuner to an antenna. You can simulate the antenna by connecting a 10-Ω resistor across the binding posts. If you use an antenna analyzer as the signal source, a 1/4-W resistor such as the RadioShack 271-1301 is suitable. But if you use your QRP transmitter, you need a total resistance of 8 to 10 Ω that will dissipate your QRP rig's output, assuming here it's 5 W or less. Four RadioShack 271-151 resistors (two series-connected pairs of two parallel-connected resistors) provide a satisfactory load if you don't transmit for extended periods. Or, you can make up your own resistor arrangement to deliver the proper load. Adjust C1 with an insulated tuning tool to achieve an SWR below 1.5:1.

Once the tuner operation is verified using the dummy antenna, it's ready to connect to the Squirt. Tuning there will be similarly sharp, and a 2:1 SWR bandwidth of about 40 kHz or so can be expected as normal.

### A Multiband Bonus

Although the Squirt was conceived with 80-meter operation in mind, it can double as a multiband antenna as well. The simple Squirt tuner is designed to match the antenna only on 80 meters. However, a good general-purpose balanced tuner such as an old Johnson Matchbox or one of the currently popular Z-match tuners (such as an Emtech ZM-2) will give good results with the Squirt on any HF band. The Squirt prototype was recently pressed into service at N2CX on 80, 40, 30, 20 and 15 meters for several months. It worked equally as well as a similar antenna fed with ladder line. Although no extensive comparative tests were done, the Squirt has delivered QRP CW contacts from coast to coast on 40, 20 and 15 meters and covers the East Coast during evening hours on 80 meters.

Build one! I'm sure you'll have fun building and using the Squirt!

### Notes

[1]Dave Benson, NN1G, and George Heron, N2APB, "The Warbler—A Simple PSK31 Transceiver for 80 Meters," *QST*, Mar 2001, pp 37-41.

[2]R. Dean Straw, N6BV, *The ARRL Antenna Book* (Newington: ARRL, 1997, 18th ed), pp 5-9 to 5-11.

[3]**www.alphalink.com.au/~parkerp/nodec97.htm; www.home.global.co.za/~tdamatta/loops.html**

[4]**home.earthlink.net/~mwattcpa/antennas.html**

[5]Full-size templates are contained in SQUIRT.ZIP available from **www.arrl.org/files/qst-binaries/**.

*You can contact Joe Everhart, N2CX, at 214 New Jersey Rd, Brooklawn, NJ 08030;* **n2cx@arrl.net**.

*Photos by the author*

QST

# Zip Cord Antennas and Feed Lines For Portable Applications

*Don't let lack of real transmission line keep you off the air!*

**William A. Parmley, KR8L**

Because of my interest in portable low power operation on the HF bands (often referred to as *HFpacking*), I decided to look into the possibility of constructing lightweight, portable antennas and feed lines using commonly available "zip cord."[1] This is a subject that was explored previously in a *QST* article that questioned the usefulness of zip cord as a feed line for the higher HF bands.[2]

Although my results are in general agreement with the earlier tests, I believe I can show that with the proper selection of material and careful deployment, feed line losses can be minimized on the higher HF bands, making zip cord dipole and feed line combinations satisfactory performers. This is especially true for portable, low-power operations in which some compromise may be acceptable in the interest of saving weight and bulk. Other potential uses for this type of antenna and feed line combination might include ARRL Field Day or a "stealth" situation in which the operator needed an easily stowed antenna that could be erected quickly whenever she wanted to operate. Of course, this would also make an ideal addition to an emergency operations *Go Kit*.

My plan was to use zip cord as a half-wavelength feed line so that the dipole's impedance would be repeated at the transmitter end of the feed line. The impedance terminating a feed line is repeated every half wavelength. Please see any edition of *The ARRL Antenna Book* for more discussion.[3] Because of this, the characteristic impedance of the feed line would be of secondary importance. Also, because the resulting feed line would be relatively short, losses would be minimized. This meant that I would not be able to erect the dipole at "optimum" height, but since the goal of this project was to create light weight, compact, pocket sized antennas and feed lines that could be used in the field with temporary supports, antenna height was not a primary consideration.

After testing a couple of different samples I settled on RadioShack No. 278-1385, #22 speaker wire as a likely candidate. This wire is sold in 100 foot rolls and has a clear insulating jacket. All electrical property measurements were made using an Autek Research Model VA1 Vector RX Antenna Analyst. Formulas and physical property information were taken from *The ARRL Antenna Book*, *The ARRL Electronics Data Book*[4] and the *VA1 Instruction Manual*. I made an effort to characterize the wire as completely as possible, as discussed below.

KR8L

**Table 1**

**Measured Velocity Factor**

| *Frequency (MHz)* | *Velocity Factor (VF)* |
|---|---|
| 3.31 | 0.68 |
| 6.75 | 0.69 |
| 13.67 | 0.70 |
| 27.77 | 0.71 |

**Table 2**

**Calculated Attenuation of Zip Cord Compared to Small Coax, dB/100 feet**

| *Frequency (MHz)* | *RS 278-1385* | *RG-174* | *RG-58* |
|---|---|---|---|
| 3.31 | 0.97 | 2.7 | 0.8 |
| 6.75 | 1.48 | 3.3 | 1.2 |
| 13.67 | 2.39 | 4.0 | 1.6 |
| 27.77 | 3.41 | 5.3 | 2.4 |

## Characteristic Impedance

This is a property that I could not measure directly with the VA1, although I was able to make some calculations and educated guesses.

To begin I measured the center-to-center distance (S) between the conductors of RadioShack No. 278-1385 speaker wire as 0.082 inches, and found the conductor diameter (d) of #22 wire listed in *The ARRL Electronics Data Book* as 0.0253 inch. For parallel conductors with air dielectric the characteristic impedance is given by:

$$Z = 276 \times \log(2S/d)$$

or

$$Z = 276 \times \log(2 \times 0.082 / 0.0253) = 224\ \Omega$$

Again, this is for an air dielectric. The plastic insulation on the wire should reduce the characteristic impedance by some amount, and although I didn't have a way to measure this effect I did come up with a way to estimate it. Here's where the "educated guess" part comes in.

First I did a similar calculation for nominal 300 Ω twin lead, yielding a result of about 400 Ω, meaning that the polyethylene insulation must reduce the calculated

[1]Notes appear on page 7.

value by a factor of about 0.75. *The ARRL Antenna Book* says that for an insulated line, the characteristic impedance can be calculated by multiplying the "in air" value by the inverse square root of the dielectric constant for the particular insulation being used. For polyethylene the dielectric constant is 2.3, so the adjustment factor should be about 0.65. However, with typical 300 Ω line the insulator is very thin so that the field between the conductors is only partially in the insulation and partially in air. It seemed reasonable that the adjustment factor (0.75 in this case) would fall somewhere between pure insulation (0.65) and pure air (1.0).

The type of insulation used on the RadioShack speaker wire isn't specified, but I made the assumption that it is polyethylene or a similar material (this seemed safe since many plastics have a similar dielectric constant of about 2.3). In addition, since the insulation between the conductors of the speaker wire is thicker (perpendicular to the plane containing the two wires) than the insulation for 300 Ω twin lead, it seemed reasonable that it must have a greater influence on the characteristic impedance since more of the field between the conductors will pass through the insulation and less through the air. Based on all of this I assumed that the adjustment factor for the characteristic impedance would probably be closer to 0.65 than to 0.75, so the characteristic impedance for this line might be:

$Z = 224 \times 0.65 = 145.6$, or about 150 Ω.

Again, this value is not particularly important, but keep it in mind and we will use it later to estimate line loss.

## Velocity Factor

The velocity factor (VF) of the line was measured using the VA1. The technique involves shorting the end of the line, then sweeping the instrument over a range of frequencies to find the lowest impedance at several points. The lowest frequency will be the frequency at which the line is a half wavelength, the next will be two half wavelengths, etc. The velocity factor can then be calculated by the ratio of the physical line length to the value of a half wavelength in vacuum at the particular frequency (given by the formula L = 492/f, where L is in feet and f is in MHz). The results for my roll of No. 278-1385 speaker wire near four amateur bands are shown in Table 1.

Since I planned to build my antennas mostly for the 20 meter band and above, I chose to use 0.70 as the velocity factor of this line. (I could just as well have said that I picked 0.70 because it is a round number or because it is approximately the average of the four readings. I think these results are amazingly consistent considering the use of a consumer grade handheld instrument. There is only a 4% variation in measured velocity factor over a frequency range of about an order of magnitude.)

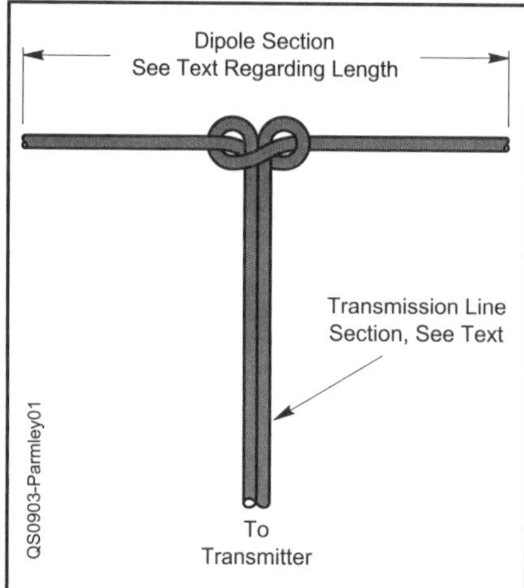


**Figure 1 — The center of the antenna section can be secured with an electrician's (or underwriter's) knot as shown.**

## Attenuation

The attenuation or line loss was also measured using the VA1. For this calculation the series of minimum impedance measurements taken during the frequency sweep (see Velocity Factor, above) were applied to the following formula from the *VA1 Instruction Manual*:

Loss = $8.69 \times Z_{MIN}/Z_0$, where $Z_0$ is the characteristic impedance of the line.

The calculated loss is given in Table 2. Values for RG-174 and RG-58 (as read from the log-log graph in *The ARRL Antenna Book*) are listed for comparison.

Now, let's think again about how I planned to use this feed line (a single half wavelength between transceiver and antenna). As the frequency increases so does the loss, but the length of a half wavelength of feed line decreases. As a result, the feed line loss remains less than 1 dB as we go up in frequency. In fact, the loss of a half wavelength line decreases from about 1 dB at 80 meters to about 0.5 dB at 10 meters.

If the estimate of a 150 Ω characteristic impedance is correct, then the SWR on the line will be about 3:1, which introduces at most an additional 0.7 dB of loss (as read from the graph of mismatched line loss in *The ARRL Antenna Book*). Total feed line loss would then be about 1.5 dB, and certainly less than 2 dB. As you can see from the table, the line is slightly more lossy than RG-58. It is closer in size and flexibility to RG-174 and is much less susceptible to damage from bending and rough handling than is coaxial cable, an important consideration given the intended use. It might be reasonable to make the feed line a full wavelength long on the higher bands in order to increase the antenna height. Whether the increase in height would offset the increase in feed line loss probably depends on the individual installation.

## Construction

Having characterized the wire as completely as possible I next wanted to check my calculations by building and testing some examples. First I calculated the antenna length using the formula $L_A = 234/f$ and then calculated the feed line length using the formula $L_F = (VF) \times 492/f$. I used the quarter wavelength formula for the dipole since the wire was going to be "unzipped" to form the two halves of the antenna and applied the standard 5% reduction in length for "end effects." I used the half wavelength formula for the feed line multiplied by the measured velocity factor. A single piece of speaker wire of length $L = L_F + L_A + X$ was then cut, where X had a couple of feet added to the length for margin of error. The feed line length ($L_F$) was measured and marked, and the remainder of the wire, including the extra length, was unzipped to make the dipole section. An electricians' knot (see Figure 1) was tied at the junction of the dipole and the feed line.

At the transmitter end of the feed line I unzipped the wire a couple of inches and attached a banana plug to one side and an alligator clip to the other. The banana plug fits perfectly in the center conductor of a transceiver's SO-239 coax connector, while the alligator clip makes a convenient way to attach to the transceiver's ground connection (as shown in Figure 2). I completely ignored the issue of feeding a balanced load with an unbalanced source here, but if you think about it, this is not an uncommon practice for simple wire antennas, and hams have been doing this kind of thing for ages. We're just more accustomed to seeing this where the feed line (coax) meets the antenna (dipole) than where the transceiver meets the feed line. For low power (QRP) applications I have not found this arrangement to be a problem. For those who are concerned about this, a few turns of the transceiver end of the feed line can be wound through an appropriate toroidal core to make a 1:1 transmission line transformer. I did test the transformer configuration briefly and found no difference in performance, feed point impedance or SWR.

For the initial setup of the dipole I measured and marked the calculated antenna length on each leg and folded the wire back on itself at this point and taped it in place. Later I found that a spring compression cord

**Figure 2 — Rear of radio showing banana plug and clip lead connections.**

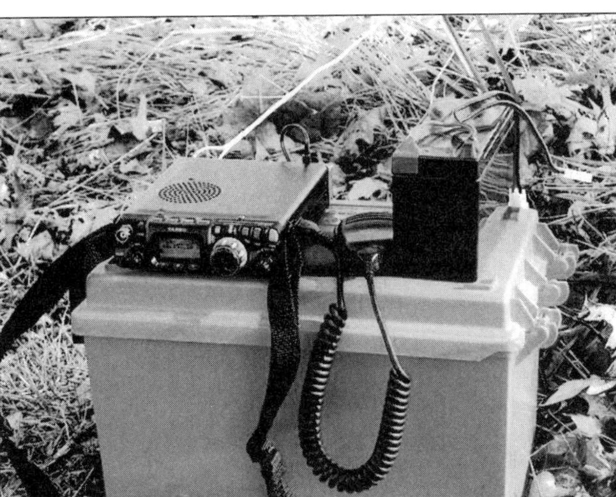

**Figure 3 — Self contained station using a zip cord antenna and feed line.**

stop of the type found on drawstrings on jackets and other items of clothing worked much better than tape, especially when the length of the dipole legs needed to be adjusted. These should be available in most fabric stores. In addition to shortening the dipole while leaving the option of lengthening it without subsequent splicing and soldering, I found that this technique created a convenient attachment loop at the end of the wire. Finally, I tied a piece of light nylon line to the loop on each end of the dipole (no other insulator was used) and proceeded with deployment and testing.

## Deployment and Testing

After building antennas and feed lines for 30, 20 and 17 meters, the initial testing was done by installing the antennas in an inverted V configuration with the apex at about 20 feet. This was done using either a telescoping fishing pole, or by tossing a line over a tree branch and pulling the dipole up with that. The ends of the dipole were brought down to 6 to 8 feet off the ground and tied off with nylon line that was then tied to tent stakes. The dipole was pruned to resonance using the VA1 by changing the fold point at the end. The extra wire was left in place and was not trimmed off. The 20 meter and 17 meter antennas were also tested as indoor dipoles by attaching the apex to a ceiling lamp and taping the ends to the walls with masking tape. In this configuration they were easily tuned to resonance.

In practice I found that once the antenna was tuned to resonance it was possible to adjust and optimize the feed point impedance by changing both the horizontal and vertical angles between the two legs. In my particular outdoor installation the best match was found with the dipole legs arranged at a horizontal (azimuthal) angle of between 90 and 120°. For indoor applications the feed point impedance was found to be adjustable by changing the amount of droop in the legs, proximity to walls or floors, and the angle between the legs. As should always be done with parallel wire feeders, I made an effort to keep the feed line clear of other objects and equidistant from both legs of the dipole to the maximum extent practicable.

## Conclusion

I have used these antennas and feed lines for SSB, CW and PSK31 at QRP power levels in indoor, backyard and backpack portable situations. Figure 3 shows a self contained portable station. Portability is excellent, deployment is simple and on-air performance seems to be very good. As noted in the original article by K1TD, some zip cord may be significantly more lossy than the type that I used, so it is important to make measurements before committing to a particular type. (One sample I tested, an example of #24 speaker wire, was at least as lossy as the wire that Jerry used.) If you want to try building your own zip cord antennas and feed lines I suggest spending some time with *The ARRL Antenna Book,* particularly the chapters on transmission lines and Smith charts. This will enhance not only your understanding but also your enjoyment of this project.

**Notes**

[1]Information about HFpacking is available on the Internet at **hfpack.com**.

[2]J. Hall, K1TD, "Zip Cord Antennas — Do They Work?" *QST,* Mar 1979, pp 31-32. This was reprinted in C. Hutchinson, K8CH, Editor, *More Wire Antenna Classics Volume 2.* Available from your ARRL dealer or the ARRL Bookstore, ARRL order no. 7709. Telephone 860-594-0355, or toll-free in the US 888-277-5289; **www.arrl.org/shop**; **pubsales@arrl.org**.

[3]R. D. Straw, Editor, *The ARRL Antenna Book,* 21st Edition. Available from your ARRL dealer or the ARRL Bookstore, ARRL order no. 9876. Telephone 860-594-0355, or toll-free in the US 888-277-5289; **www.arrl.org/shop**; **pubsales@arrl.org**.

[4]*The ARRL Electronics Data Book* is out of print.

*Photos by the author.*

*Bill Parmley, KR8L, was first licensed in high school, then dropped out of Amateur Radio while in college and the military. He was licensed again in 1979 and upgraded to Amateur Extra in 1981. He is an ARRL member. His interests range from QRP on 160 meters to meteor scatter on 222 MHz and from CW to digital voice. Bill has a Master's degree in physics from Michigan State University. He served as a nuclear submarine officer in the US Navy, and has worked as a nuclear engineer for several electric utilities, and as a safety engineer, project manager and program manager for the US Department of Energy. Bill and his wife Anne, KA8TER, are now retired and enjoy living on the small southern Illinois farm where he grew up. You can read more about his Amateur Radio interests and activities at* **kr8l.us**, *and you can contact him at 1123 Country Club Rd, Metropolis, IL 62960 or at* **kr8l@kr8l.us**.

# The Octopus — Four Band HF Antenna for Portable Use

*An easy to make, easy to assemble and easy to pack travel antenna.*

**Geoff Haines, N1GY**

My participation in the world of Amateur Radio has led me in many directions. From DX to disaster recovery, I have used my meager skills to do many things. For me, the most enjoyable of these is the building of the simple gadgets and accessories that make my station work better.

## One for the Road

I have a frequent need for an HF antenna to support portable operation on multiple bands. I recently thought of a simple way to provide one based on a *multiband fan dipole* in combination with a *helical mobile dipole*. Now those of you with 70 foot towers with multi-band Yagis on top who never leave their linear behind can leave the room. This idea is for those who like to go somewhere such as a beach or park where it's quiet and set up a modest station for an afternoon of radio fun. It also works for those of us who are Amateur Radio Emergency Service (ARES) members and have to set up a working HF station in the middle of a parking lot after a natural disaster such as a hurricane.

## The Octopus is Born

Using the concept of a horizontal fan dipole, I drew up a design for a helical type dipole taken to extremes. The basic plan was to find an attachment mechanism that could be mounted on a mast and accept up to four helical dipoles, eight helical type antennas, radially around the mount in the horizontal plane.

**Figure 1 — The hub of the array is a standard octagonal electrical box available at your local home improvement store. Stud mounts available from your local ham radio or CB shop are used to mount the mobile antenna pairs.**

### *Finding the Pieces*

Since I have often found the perfect parts for my projects at the local home improvement chain stores, I looked there again. Almost as soon as I got to the electrical department, my search was over. An eight sided electrical box, of the type used to mount overhead lighting, looked ideally suited for the task. See Figure 1. It had eight sides, a top and a bottom and would be relatively easy to modify. The punch-out slugs on the side were larger than I wanted for mounting the helicals, but could be adapted with fender washers. I drilled these out to ½ inch internal clearance.

The punch-out slugs in the top and bottom of the box are the right size "out of the box" to allow the entire mount to be slipped over the top of a ¾ inch diameter top section of telescoping mast. I used an angle bracket bolted to the bottom of the box to allow me to clamp it to the mast.

The next problem was how to mount the antennas. Checking the hardware aisle I found ⅜ × 24 bolts in the fasteners section. The long nuts for the other side of the mount were harder to find, but by going to truck stops (where they usually have a CB parts wall, visiting RadioShack and raiding my parts box, I eventually came up with eight double female stud mounts to attach to the central hub. The assembly details are shown in Figure 2.

### *Making it Happen*

Four of these stud mounts were mounted directly to the hub, and four were mounted with insulating washers. The four insulated mounts were wired to a single SO-239 type connector mounted to the bottom of the hub box. Each insulated mount has a grounded mount 180° away from it. The connections from the insulated mounts to the SO-239 were made with ring terminals of appropriate size soldered to a wire harness that is connected to the SO-239, as shown in Figure 3.

Since this antenna system is intended for short term, temporary deployment, weatherproofing is not particularly significant for me. Should anyone want to use this idea for a more permanent installation, weatherproof-

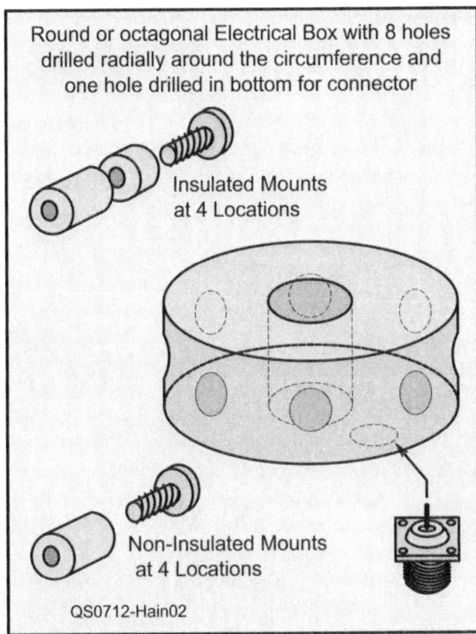


**Figure 2 — A drawing of the electrical box showing the holes drilled for the stud mounts, four of which are insulated from the box and four of which are grounded to it.**

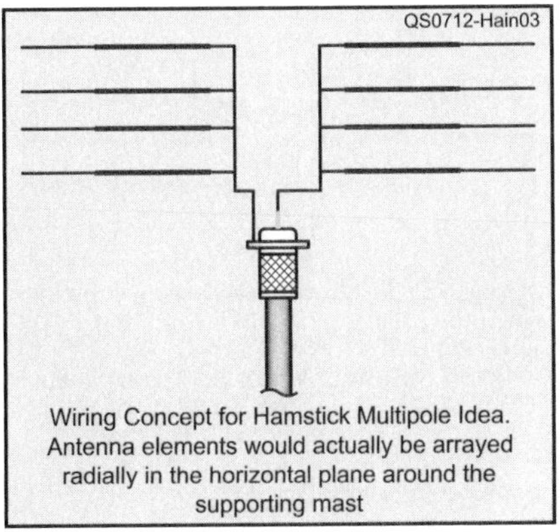

QS0712-Hain03
Wiring Concept for Hamstick Multipole Idea. Antenna elements would actually be arrayed radially in the horizontal plane around the supporting mast


**Figure 3 — This shows the schematic diagram of the wiring of the antenna elements.**

**Figure 4 — Here is a view of the octopus all set up for park-side DX or disaster recovery operations. The SUV, sitting on the mast base, keeps the whole array erect. The guy lines in the picture would be used only if the vehicle was needed elsewhere.**

ing the hub will require at least several coats of rust resistant paint and the sealing of any spots where water could gain entry.

The next step was to install four pairs of helical type mobile antennas cut for four different bands on the mount. With each pair, one is connected to an insulated mount and the other is installed in the grounded mount opposite to the first. I chose to use Valor Pro-Am helical whip antennas because the 3/8 × 24 connection of the whip to the helically wound lower shaft of the antenna makes them easier to store and assemble than the set-screw arrangement of the Lakeside Hamstick antennas. This is not to say one is better than the other. I have purchased both makes before and indeed use the Lakeside antennas on my mobile rig where storage is not an issue. The antenna, all ready to go, is shown in Figure 4.

## How it Works

Since we now have four dipoles all connected to the same coax, a little explanation is in order. When the radio transmits RF up the coax to the antenna array, it sees all of the antennas as a high impedance load except the dipole that is resonant on the band being transmitted. Thus, almost all of the RF energy is directed automatically to the proper dipole. This is the way a multiband fan dipole works, and what we have here is such an antenna turned sideways from the usual configuration. See Figure 5.

## Tuning it Up

There are two ways to make sure that the transceiver sees a 50 Ω load. One is to mount and raise one pair of antennas at a time, and tune each pair to present a 50 Ω load. It's very time-consuming, but it does work. Each pair will have to be retuned to some extent because of slight interactions between the pairs of antennas, but eventually all can be made to work. The second way is to get them approximately on frequency and use a manual or automatic antenna coupler or tuner such as my LDG Z-100.[1] This was my personal choice.

Once the antenna array goes up, I don't want to have to take it down until I am ready to go home. I chose to adjust each dipole to resonance at the middle of each band. To expand the bandwidth to the full scope of the available frequencies I rely on the automatic tuner, which also appears to handle the small mismatches caused by any interaction between sets of dipoles. This interaction is less than with a traditional fan dipole because the 40 and 75 meter antennas are at right angles to each other. The 20 and 15 meter antenna sets are also at right angles to each other and at a 45° angle to the lower band sets.

[1]Notes appear on page 10.

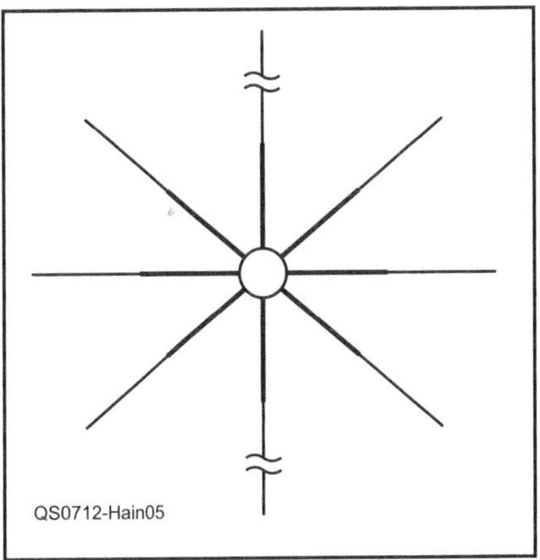


**Figure 5 — Top view of the array.**

## Raising the Beast

The octopus array is mounted on a telescopic mast home-brewed of aluminum tubing available from several sources. The telescopic feature allows the array to be positioned from a low position for near vertical incidence skywave (NVIS) type operations or extended up to 26 feet for lower angle operation on the higher bands.[2] Any height up to λ/4 will support NVIS operation, so the full height works well on 80 and 40 meters; the prime NVIS bands but lower heights can also be effectively used, if conditions warrant. I think that the low position should be above 8 or 9 feet off the ground for safety's sake. No one likes to run into the sharp end of an antenna.

## Don't Expect Miracles — but Good Results for its Size

Any short dipole array, including the octopus, is a compromise. If time were no object and funds unlimited, there are antenna systems that work better. However, if your antenna system has to ride around in the trunk of your car waiting for the next ARES callout, or you just have one afternoon to devote to catching some fun DX at the beach, this antenna will do the job well enough. A decent enough antenna is better than no antenna at all.

To store the array, just remove the helicals and put them back into whatever bag they are normally stored in. The mast is telescoped to its minimum length and stored the same way. You can leave the hub on the mast or store it separately — your choice. The initial setup only takes about 15 to 20 minutes and once done, you are finished with antenna work until it's time to go home. There are no radials to lay out and trip over. When disassembled, it all fits in the trunk of your car.

## But Wait — There's More!

This antenna array also allows the use of a VHF or UHF antenna mounted above the hub on the very tip of the mast as shown in the lead photo. Therefore, with just one mast you can provide communications over a broad range of amateur bands and ranges with a minimum of effort. Sounds like the Amateur Radio we all know and love.

## How's it Play?

I raised the octopus about 11 feet above ground for my initial testing. The radio and autotuner combination had no trouble getting a good match on any of the four bands, 75, 40, 20 and 15 meters. During an all-day display of Amateur Radio at a local American Red Cross center, I set up the mast and antenna

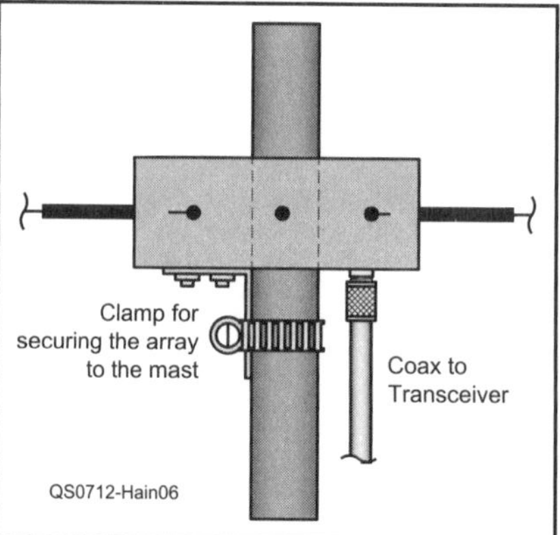


**Figure 6 — This diagram shows clearly how the hub is secured to the mast to allow "armstrong" rotation of the array, if needed.**

to its full height of 26 feet. I had no trouble at all checking into the Maritime Mobile Service Net on 20 meters and similar success in making contacts on all of the bands except 75 meters, which is usually pretty dead here during the daytime. I was able to get a good match on all four bands, so I have no doubt that it will work on 75 meters when the band is open. To take advantage of the directional effects of the dipoles, particularly on the higher bands, the antenna can be manually rotated as shown in Figure 6.

One cautionary note: This antenna mast should never be tilted up with the antenna on board. It should always be extended upward with the mast already vertical. I discovered this when I attempted the tilt-up method, with disastrous results. The mast bent under the weight of the antenna and had to be repaired. As long as it is extended vertically, and then guyed, there should be no problem in any reasonable wind.

If you decide to try this approach, you can substitute your favorite band or bands of choice. You could build a one, two or three band version just as easily. In terms of overall performance, the array appears to work just as well as any single band helical type dipole. Note that any helical dipole will be more efficient at 20 meters and higher than on the lower bands, but will work as well as other antennas their size. The main advantage of the multiband configuration is that when you want to change bands you just switch the radio to the new band and give the TUNE button a jab. Two seconds later you are ready. No going out to the mast, lowering it down to 6 feet and removing one set of antennas to substitute another pair. All of that could easily soak up 15 or 20 minutes, and you still have to raise the mast back up to the proper height.

## Time to Hit the Road

The octopus resides in the back of my SUV, along with the telescopic mast and its base, also home-brewed. When deployed, the base sits under one of the tires on the SUV and keeps the mast vertical until the guy ropes are deployed. Then I can drive the car away and set up the rest of the station in a more convenient spot. Total time from the initial setup of the mast and antenna to being on the air ready to catch some DX or assist after a disaster is about 30 minutes. It can be less if I use the radio mounted in the car instead of a separate station set up as in the "Radio in a Box" pictured in *Up Front* in the October 2005 issue of *QST*. On the other hand, using the radio in a box means that the vehicle is available to do other things while the station continues to be operated.

### Notes

[1]J. Hallas, W1ZR, "Product Review — Automatic Antenna Tuners — A Sample of the Field," *QST*, May 2004, pp 71-76.

[2]R. Straw, N6BV, "What's the Deal About 'NVIS'?" *QST*, Dec 2005, pp 38-43.

*Geoff Haines, N1GY, has been licensed since 1992 and holds an Amateur Extra class license. He retired after a career in respiratory care. He currently holds several ARRL appointments in the West Central Florida Section, including Technical Coordinator, Technical Specialist, Official Bulletin Station, Net Manager, Official Emergency Station and Official Relay Station. He is the President of the Manatee Amateur Radio Club, a member of the Manatee ARES group and member of the Bradenton Amateur Radio Club, the Yale University Amateur Radio Club and the Meriden (CT) Amateur Radio Club. In his spare time, he enjoys homebrewing antennas and accessories for his Amateur Radio operations. Geoff can be reached at 708 52nd Av Ln W, Bradenton, FL 34207;* **n1gy@arrl.net**.

# 8-Band Backpacker Special

## Here's a portable antenna that you can use for backpacking or camping—not to mention Field Day.

**By Jim Andera, WBØKRX**
506 S Center St
Gardner, KS 66030

Many hams have discovered the value of using Amateur Radio in conjunction with other hobbies ranging from backpacking to bicycling. Those convenient VHF or UHF hand-held radios are great for short-range communications, but what if you need to talk hundreds or thousands of miles? The answer: HF radios, often QRP style to minimize size and weight. A problem you'll run into as you get set up for portable HF operation is finding an easily transportable antenna that's efficient enough to let your QRP signal reach out and touch someone.

A simple dipole antenna, usually strung inverted-**V** style, is hard to beat in terms of efficiency and simplicity. Fan-dipoles work well for two or three bands, but how

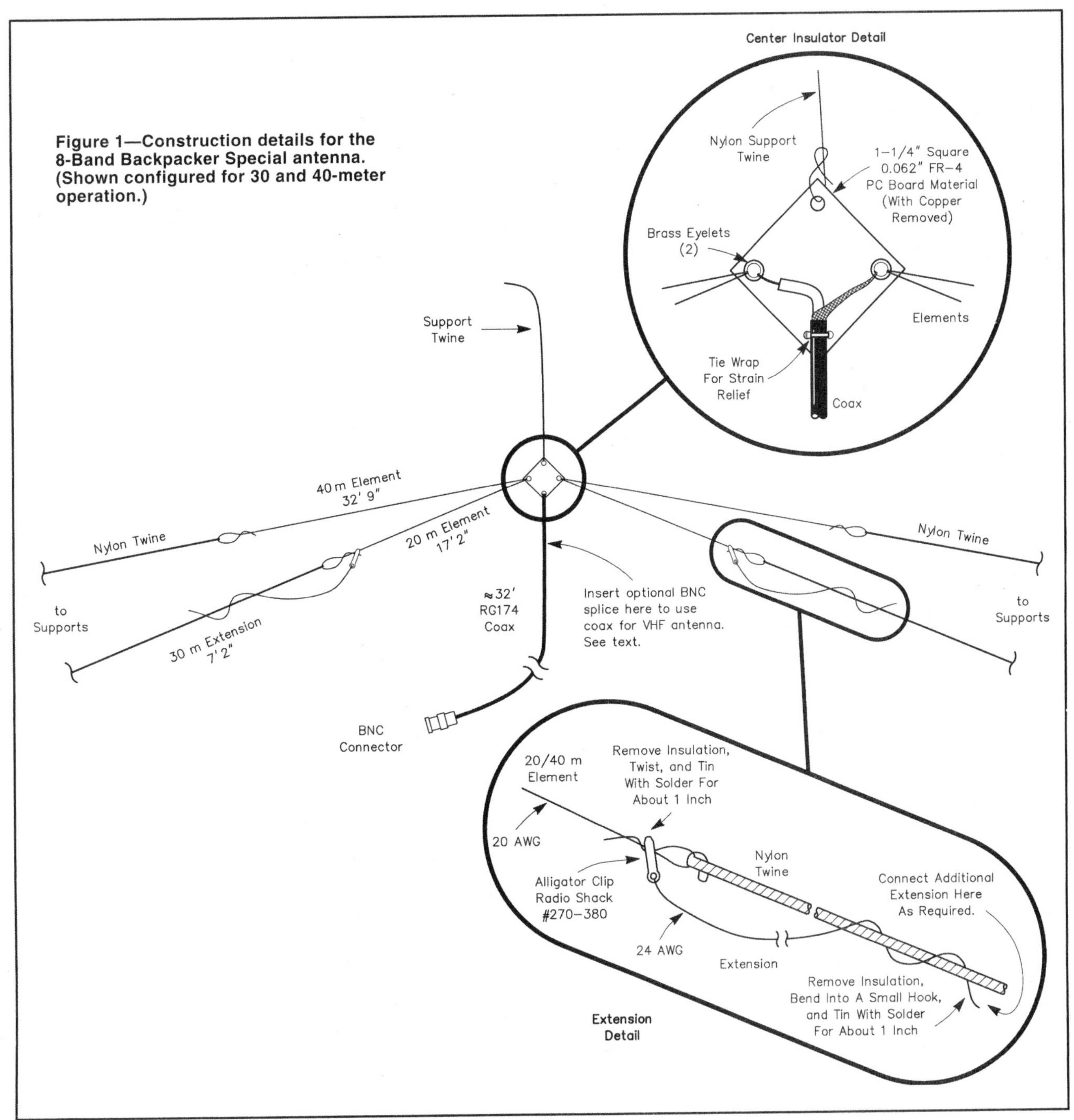


**Figure 1—Construction details for the 8-Band Backpacker Special antenna. (Shown configured for 30 and 40-meter operation.)**

**Table 1**

**Eight-band operation can be achieved by extending a 20-meter/40-meter fan dipole. Extensions are clipped to the end of the basic 20- or 40-meter elements.* Carrying an additional pair of 6-inch extensions allows for fine-tuning the antenna.**

| *Band (CW)* | *Basic Element* | *Extension Length†* | *Total Element* Length* | *Antenna Electrical Length* |
|---|---|---|---|---|
| 80 m | 40 m | 32 ft 11 in | 63 ft 8 in | $^1/_2\ \lambda$ |
| 40 m | 40 m | none | 32 ft 9 in | $^1/_2\ \lambda$ |
| 30 m | 20 m | 7 ft 2 in | 24 ft 4 in | $^1/_2\ \lambda$ |
| 20 m | 20 m | none | 17 ft 2 in | $^1/_2\ \lambda$ |
| 17 m | 40 m | 7 ft 2 in | 39 ft 11 in | $1^1/_2\ \lambda$ |
| 15 m | 40 m | 1 ft 6 in | 34 ft 3 in | $1^1/_2\ \lambda$ |
| 12 m | 20 m | 7 ft 2 in and 4 ft 9 in | 29 ft 1 in | $1^1/_2\ \lambda$ |
| 10 m | 20 m | 7 ft 2 in and 1 ft 6 in | 25 ft 10 in | $1^1/_2\ \lambda$ |

*An element is one-half of a dipole.
†Two extensions are required, one for each element.

would you make an easily transportable antenna to cover 80 through 10 meters without creating a tangled mess of wires? Try this 8-Band Backpacker Special! This is what I use on my backpacking adventures; it lets me hop between two bands without even having to get out of my sleeping bag to make adjustments. (See my article, "Operating Backpack Portable," *QST,* April 1994.)

### Construction

The Backpacker Special consists of a 20-meter/40-meter fan dipole that has extensions clipped to the ends of the elements to make the overall antenna $^1/_2$ or $1^1/_2$ wavelengths long on each band. (As used in this article, an *element* is one-half of a dipole.) Table 1 shows the lengths of the various extensions that need to be added to the basic antenna to make it cover the eight HF bands; each extension has an alligator clip soldered to one end. Several extensions are reused in the various configurations to minimize the amount of wire carried, reducing the weight and bulk of the antenna. (My pack dog, who carries the antenna on backpacking trips, appreciates the fact that the antenna weighs only 1.4 pounds and doesn't take up much space—it leaves more room for dog chow.) The lengths given in Table 1 typically resonate the antenna on the low edge of the CW bands; for the Novice subband the length of the 40-meter element may need to be shortened about 4 inches, and the 80-meter extension shortened about 18 inches.

Figure 1 illustrates the details of the antenna's construction. I like to use #20 AWG magnet wire for the main elements and #24 AWG for the shorter extensions. The #20 AWG is strong enough to allow it to be pulled down when tangled in tree branches, without breaking the wire. About 32 feet of RG-174 coax is convenient for the feed line. Make sure to strain-relieve the coax at the center insulator. A BNC splice can be inserted close to the center insulator if you want to be able to steal the coax to use with your VHF antenna (a 30-foot length of RG-174 will have about 3 dB loss at 144 MHz).

Check the antenna with an SWR meter before you take off on your adventure by setting it up in the backyard or in a park. Install it at whatever apex and element height you think you'll usually use. When I use it, the apex is often supported by a tree branch 15 to 25 feet above the ground and the ends of the elements are 6 to 8 feet high. If you let the ends of the elements get too high, it's hard to reach them to attach the extensions!

### Materials

You may be able to use your creativity and junk box to reduce the need to go out and buy everything for this antenna. Few, if any, ham dealers will carry the miniature RG-174 coax or connectors. RG-58 is fine if you don't mind the extra bulk and weight. The nice thing about the magnet wire I've specified is that it has a thin film of insulation that will burn off easily during soldering (no need to strip the insulation in advance); but most any type of wire will do. FR-4 PC board material can be identified by viewing the edge of the board and looking for thread-like filaments imbedded inside the material. If you don't have any old FR-4 in your junk box, substitute a high-temperature plastic or some varnished wood (avoid phenolic PC board material because it's too brittle).

If you need a source for the wire and connectors try your local electronics distributors, or you can order them by telephone from:

Arrow/Capstone: 800-833-3557 ($50 minimum order)

Newark Electronics, either your local outlet or their national order line: 800-281-4320 ($25 minimum order)

Parts list:

#20 AWG magnet wire—Belden 8050 (approx. 160 ft/0.5 lb spool)

#24 AWG magnet wire—Belden 8052 (approx. 404 ft/0.5 lb spool)

RG-174 miniature coax—Belden 8216 (it comes in a 100-foot spool)

BNC Plug for RG-174 (clamp style)—Amphenol type 69475

BNC Plug for RG-174 (crimp style*)—Amphenol type 31-315

BNC Jack for RG-174 (crimp style*)—Amphenol type 31-317

Hardware stores and Radio Shack have various sizes and styles of alligator clips, tie wraps, and brass eyelets that are useful.

*Crimp connectors can be used without a crimp tool by soldering the collar over the coax braid rather than crimping it on. Take care not to melt the coax center conductor. After it's soldered, place a couple of inches of heat-shrink tubing over the collar and coax jacket for strain relief.

### Antenna Tips

❑ If you mount the antenna unusually low, the antenna will behave as though it's longer and tune up low in frequency. Clipping the extensions on a few inches from the end of the basic element will help compensate for this effect. If mounted high, the antenna will act as if it's shorter, so you may need to add a short extension (try six inches for starters) to make it tune properly.

❑ Ground conductivity can affect the tuning. Raising or lowering the ends can help compensate for this variation, as can the two remedies mentioned above.

❑ Nylon twine works well as an end insulator and support. Have these ropes pre-cut (the apex rope 50 feet long and the element ropes 20 feet long) and wind them up on some type of form. A few 3 and 6-foot sections are handy, too.

❑ Keep the entire kit in a gallon-size Ziplok plastic bag to keep things together, with smaller items in a quart-size bag. QST

By Robert H. Johns, W3JIP

# Build a Portable Antenna

Can't work daytime skip? Take your rig on vacation! This HF/VHF antenna mounts anywhere, packs well and works like a champ.

This dipole antenna can be used for any band from 80 through 2 meters. One half of the dipole is an inductively loaded aluminum tube. Its length is adjustable from 4 to 11 feet, depending on how much room is available. The other half is flexible insulated wire that can be spooled out as necessary. The tube is supported by a flagpole bracket attached to a long carpenter's clamp. The clamp mounts the antenna to practically anything: a windowsill, railing, a chair or post. If there is no structure to mount the antenna (a parking lot or the beach), the clamp attaches to two light wooden legs to form a tripod.

The key to mounting flexibility is the large clamp. The key to electrical flexibility is a large adjustable coil that lets you resonate the tube on many bands. The coil is wound with #8 aluminum ground wire from Radio Shack. The form is a four-inch (4 1/2-inch OD) styrene drain coupling from The Home Depot or a large plumbing supply store. A movable tap adjusts the inductance to tune the upright tube to the desired band. The wire half of the antenna is always a bit less than $\lambda/4$ on each band. Hang it from any convenient support or drape it over bushes to keep it off the ground.

## Construction

The 18-inch carpenter's clamp (sometimes called a bar clamp, such as Jorgensen's No. 3718) and flagpole bracket that takes a 3/4-inch pole are common hardware items. The bracket is insulated from the clamp jaw by a 1 1/2-inch length of 1-inch PVC pipe (see Figure 2). Hammer the PVC over the end clamp jaw to make it take the shape of the jaw. Secure the flagpole bracket to PVC with a large ground clamp (for 1/2 or 1-inch conduit). The ground clamp includes 1/4-inch bolts; enlarge the flagpole bracket holes to accept them. Some flagpole brackets have an integral cleat; I hammered the cleat ears flat on mine.

Figure 1—The portable antenna in some of the many places it may be mounted around the house, porch and yard.

Mount an SO-239 chassis connector on the flagpole bracket using Radio Shack insulated standoffs (276-1381). The standoffs tightly press the center pin of the SO-239 against the bracket surface; no other connection is needed. Other mounting hardware may require a connection from the coax center conductor to the bracket. The spooled wire's inner end wraps around a standoff and connects to a ring terminal under a screw holding the SO-239 flange to the standoff.

The 1×2-inch wooden legs for the tripod are each 30 inches long. Bolt them together at one end with a 1 or 1 1/4-inch-long bolt. Countersink the bolt head and nut below the surface of the wood so they don't interfere with the clamp jaws.

## Aluminum Element

This element can be made from three lengths of telescoping aluminum tubing (3/4, 5/8 and 1/2 inch, 0.058-inch walls). I used tubing with thinner walls for less weight and easier handling. A 45-inch-long, 3/4-inch tube fits the flagpole bracket. I chose this length because it and the flagpole bracket make $\lambda/4$ on 6 meters. The two outer tubes are both 5/8 inch, made by cutting a seven-foot aluminum clothesline pole in halves. They are joined together with a copper coupling (5/8 inch ID) for 1/2-inch copper pipe. The coupling is bolted to one of the 5/8-inch sections, and a slot is cut in the free half of the coupling. The

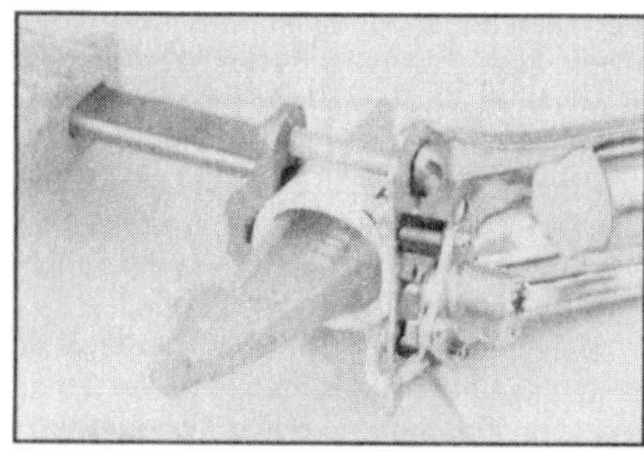

Figure 2—The flagpole bracket that supports the tubing elements is clamped to the long carpenter's clamp, but insulated from it by a small section of 1-inch PVC pipe. A coax connector is mounted to the bracket and the spool of wire is attached to the coax connector.

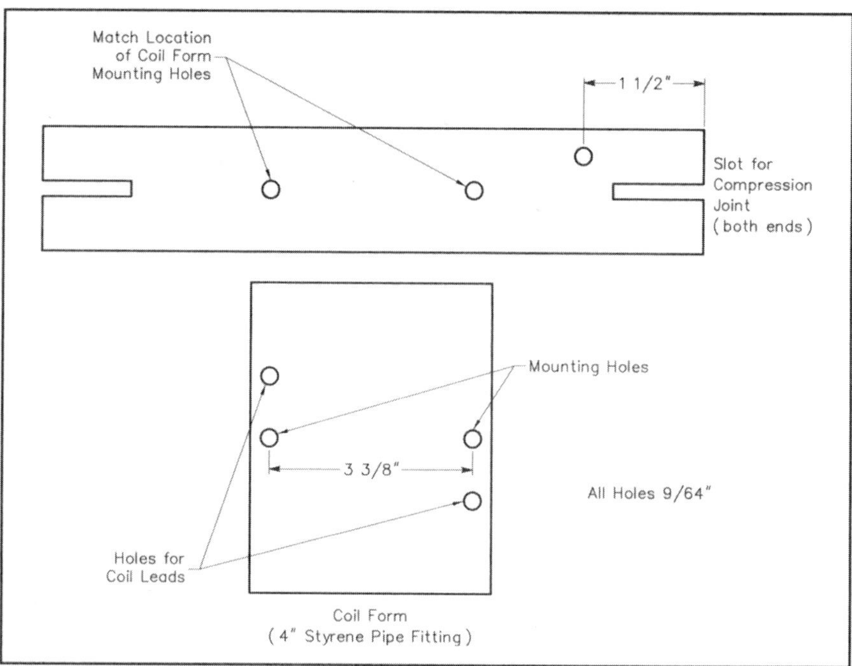


Figure 4—Holes to be drilled in the styrene coupling and the 3/4-inch PVC pipe. All holes are 9/64-inch diameter, to provide clearance for #6-32 bolts. The hole 1 1/2 inches from one end holds a bolt that serves as a stop, so that the antenna tube does not slide in too far. Space the holes for coil leads far enough from the mounting holes to clear the 3/4-inch pipe.

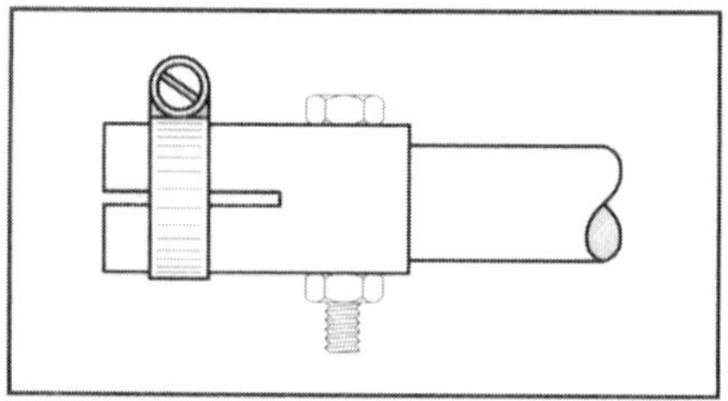

Figure 3—The joint between two sections of 5/8-inch tubing is made from a 1/2-inch copper pipe coupling, bolted to one section and hose clamped to the other.

remaining 5/8-inch tube is inserted there and secured with a hose clamp. See Figure 3.

One 5/8-inch tube slides into the 3/4-inch tube to provide a continuously variable element length from about 4 to 7 1/2 feet. This extends from 7 1/2 to 11 feet when the two 5/8-inch sections are joined together.

## Loading Coil

The loading coil has 12 turns of bare aluminum wire spaced to fill the 3 1/2-inch length of the drain coupling. Drill 9/64-inch holes at the ends of the coil form to accept the ends of the coil wire. (See Figure 4.) To wind the coil, feed four inches of wire through the form, make a sharp bend in the wire and start winding away from the nearby mounting hole. Wind 12 turns on the form, spacing them only approximately. Cut the wire for a 4-inch lead and feed it through the other hole in the form. Tighten the wire as best you can and bend it into another acute angle where it passes into the form. Space the turns about equally, but don't fuss with them. Final spacing will be set after the wire is tightened.

Tighten the coil wire by putting a screwdriver or a needle-nose pliers jaw under one turn, and pry the wire up and away from the surface of the coil form. This can be done anywhere but I prefer these kinks on the backside, away from the mounting shaft. Put kinks in every other turn. This removes any slack from the coil and holds the turns in place. Should the coil ever loosen, simply retighten it with a screwdriver. If you prefer, glue the coil turns in place with epoxy or coil dope. Use a thin bead of glue that won't interfere with the clip that connects to the coil.

The coil form mounts on a nine-inch-long 3/4-inch PVC pipe.[1] (See Figures 4 and 5.) The inside diameter is slightly larger than the 3/4-inch aluminum, but slotting the PVC and tightening it with a hose clamp secures the tube. (Use a wide saw to cut these slots, not a hacksaw.) The coil form mounts to the PVC pipe with #6-32× 1 1/2-inch bolts. A five-inch long, 3/4-inch aluminum tube permanently attaches to one end of the coil assembly and slides into the flagpole bracket. One end of the coil wire connects to this short tube. Flatten the wire end by hammering on something hard, then drill a 9/64-inch hole in the flattened end and attach it to the short tube with a #6-32×1-inch bolt. Tighten the bolt until the 1/2-inch tube starts to flatten. This will keep pressure on the aluminum-to-aluminum joint.

The aluminum element slides into the opposite end of the coil assembly. The hose clamp there can be tightened until the element just slides in snugly.

A 12-inch clip lead connects the aluminum element to the coil. Bolt the plain wire end to the 3/4-inch tube three inches from its end. Many alligator clips will fit in the space between the turns of the coil (about 3/16 inch), but I prefer using a solid-copper clip (Mueller TC-1). I cut off most of the jaws, so that only the part close to the hinge grabs the coil. This shorter lever grips very tightly.

## Wire and Spool

The lower half of the antenna is insulated wire that is about $\lambda/4$ on the band of operation. The wire is pulled from the spool, and the remaining wire forms an inductance that doesn't add much to the antenna length. The Home Depot sells #12 and #14 insulated stranded copper wire in 50 and 100 foot lengths, on plastic spools. (See Figure 6.) A 1/2-inch dowel fits into the spool to become a handle and spool axle. Bolts through the dowel on either side of the spool hold it in place. A crank handle is made by putting a one-inch-long bolt through the spool flange.

I calibrate the wire on the spool with markers and electrical-tape flags. There's a mark (from a permanent marking pen) at each foot, a black flag every five feet and a length-marked colored flag every 10 feet.

[1]The PVC pipe that is generally available, Schedule 40, is okay, but a stronger support can be made with 3/4-inch Schedule 80 PVC pipe. It has the same outside diameter with a thicker wall. It can be made even stronger by forcing a 3/4-inch wood dowel into it. You must drill or file the inside of the pipe to fit 3/4-inch aluminum tubing into it. Schedule 80 PVC pipe is available from large plumbing distributors. Call before shopping. An even more rigid coil support can be made with a 1-inch OD, 1/8-inch wall fiberglass tube, available from Max-Gain Systems, Inc, 221 Greencrest Ct, Marietta, GA 30068-3825; tel 770-973-6251.

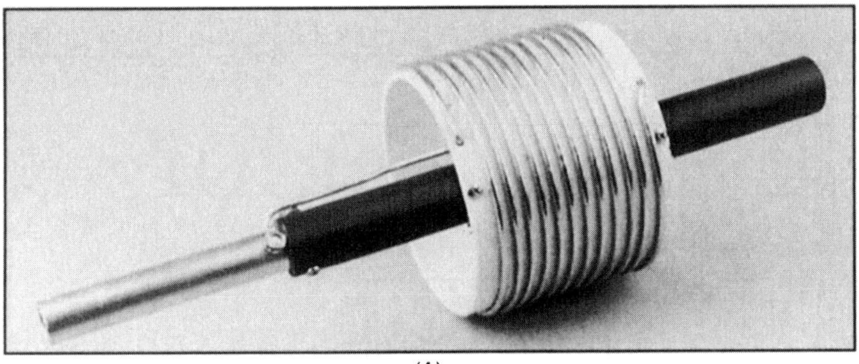

(A)

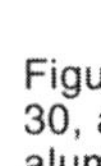

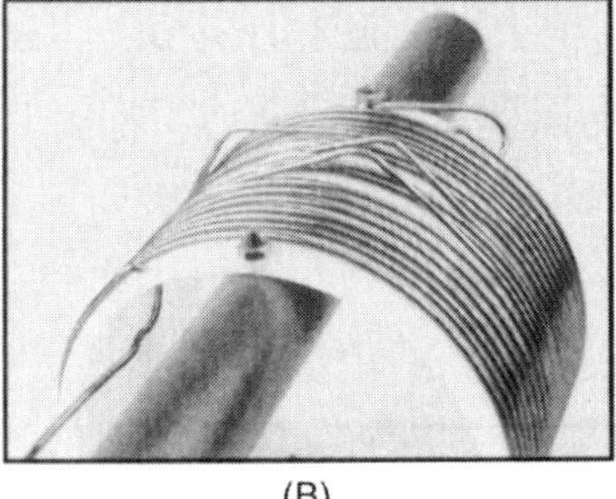

(B)

Figure 5—A shows the loading coil for the 20, 30, and 40-meter band coverage. The short aluminum tube on the coil slides into the flagpole bracket, and the tubing element slides into the other end of the PVC pipe. The wire and clip connect the element to the coil. B shows the coil for operating the antenna on 80 meters. This is placed in the flagpole bracket and the 40-meter aluminum coil plus the tubing element is inserted into it.

**Table 1**

| | *Tubing Length (ft)* | *Coil Turns* | *Wire Length (ft)* |
|---|---|---|---|
| **6 meters** | 4 | 0 | 4 |
| **10 meters** | 7 | 0 | 8 |
| | 4 | 1.7 | 7 |
| **12 meters** | 8 | 0 | 8.3 |
| | 4 | 1.8 | 9 |
| **15 meters** | 10 | 0 | 10.3 |
| | 8 | 1.8 | 10 |
| | 4 | 3.5 | 10 |
| **17 meters** | 11 | 0 | 13 |
| | 8 | 2.2 | 13 |
| | 4 | 4.2 | 13 |
| **20 meters** | 11 | 1 | 16 |
| | 8 | 4 | 16 |
| | 4 | 6.2 | 16 |
| **30 meters** | 11 | 4 | 24 |
| | 8 | 7 | 24 |
| | 4 | 10.5 | 24 |
| **40 meters** | 11 | 9 | 32 |
| | 8 | 12 | 32 |

Simply mark the length for each band, if you like.

We must prevent the wire unspooling, especially when it's hanging from a window mounting. A heavy rubber band works, but it doesn't last long. A better solution is a loop of light bungee cord, preferably with a knot for grip. The bungee loop runs from the axle/handle around the spool making a half twist on the way, and then passes over the axle end on the other side of the spool. (See Figure 6.)

## Operation

Table 1 lists element length, wire length and coil tap point for various bands. When the "turns" is zero, the coil is not needed.

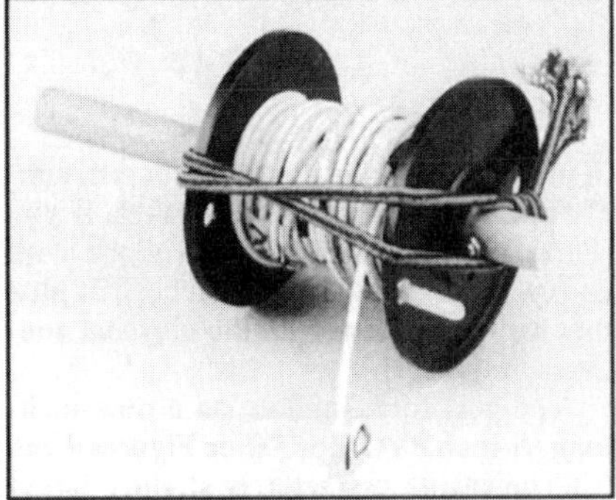

Figure 6—The wire spool has a wooden axle/handle and small handle for winding the wire. A bungee cord stretched over the spool and around the axle prevents the wire from unwinding.

On all bands except 6 meters, you can simply bypass the coil with the clip lead, the extra length just lowers the frequency a bit. For 6 meters, the coil *must* be removed. The location of the unspooled wire greatly affects the settings, so these numbers are only starting points. The lengths in the table were taken with the wire one to three feet above ground, draped over and through bushes and flowerbeds. The antenna will still work if the wire is lying on the ground, but it will require less unspooled wire to resonate. An SWR analyzer is very helpful while adjusting this antenna. I have not used a balun with this antenna (except for some testing), and I don't think one is necessary.

The SWR is less than 1.5:1 on all bands, and it's usually below 1.2. Occasionally, a band shows a higher SWR (still less than 2:1), but I have always been able to lower it by adjusting the length or location of the lower wire. I have not used the antenna from a high window where the wire hangs straight down, so cannot say what the SWR will be in this case. Never set up the antenna where it could fall and injure someone, or where the unwary could get an RF burn by touching it.

My results with this antenna have been excellent, both from home and on vacation. If you haven't yet operated from a seashore location, be prepared for a pleasant surprise! The good ground afforded by the salt water really makes a difference. In addition, I believe that coil radiation contributes to the success of the antenna.

## 80 Meters

It's easy to add this lower frequency band. Figure 5B shows a 35 μH coil for 80 meters. It's constructed and tightened just like the 40-meter coil, but has 20 turns of #12 magnet wire.

To operate on 75/80 meters, insert the new coil into the flagpole bracket and plug the 40-meter coil into it. Tune across the band with the movable tap on the 40-meter coil. This varies antenna resonance from below 3.5 to above 4.0 MHz, with the full 11 feet of tubing extended. If your version doesn't achieve this tuning range, adjust the spacing of the turns on the 80-meter coil.

The 80-meter coil has a five-inch length of ³/₄-inch aluminum tube inserted into one end of the ¹/₂-inch PVC pipe that supports the coil form. One end of the coil is connected to this aluminum tube. The other end is secured under the bolt that holds the coil form to the PVC pipe. A second clip lead connects the base of the 40-meter coil to the outer end of the larger coil. The length of wire on the spool must also be increased to about 64 feet.

## 2 Meters

A $\lambda/2$ dipole for 2 meters can be made with about 15 inches of ¹/₂-inch aluminum tubing in the flagpole bracket, and 18 inches of wire. The tubing element is shorter than normal for 2 meters because the bracket is also part of the antenna. You can also shorten the 6-meter wire a bit and operate the 6-meter antenna as a $3\lambda/2$ on 2 meters, with a somewhat higher SWR.

## Continuous Coverage

With easily changed element lengths and a continuously variable loading coil, the antenna may be operated on any frequency from 6.5 to 60 MHz, if coverage for other services is needed. With taps in the 80-meter coil at 8, 11 and 14 turns, the antenna will also tune from 4 to 7 MHz.

*Bob Johns is a semiretired physics teacher. First licensed in 1952, he builds and experiments with coils, traps and antennas. He can be reached at Box 662, Bryn Athyn, PA 19009.*

QST

By Dennis Kennedy, N8GGI

# A Packable Antenna for 80 through 2 Meters

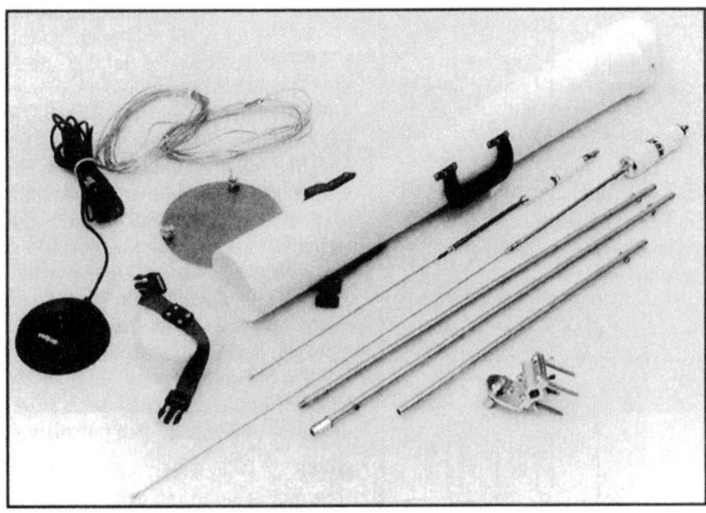

N8GGI converts a Hustler mobile antenna into a versatile airline-shippable antenna for mobile, marine or portable operation.

Here are 40, 20, 6 and 2 meter antennas, two radials for 20 meters, mounts and case.

The growing popularity of pint-sized HF and HF/VHF transceivers convinced me that I need one. I use it as a mobile rig, on business trips, vacations and to activate some US Islands. While the radio and power supply now travel easily, the antenna was the limiting factor for travel.

I bought a Hustler mobile antenna mast and 20 meter resonator and began using them with good success on the car. The fold-over 54 inch mast doesn't suit all the possible operating situations my wife might allow because it only folds to 90°. I wanted a versatile, compact antenna system that could pack compactly enough to ship as airline baggage. It would have a segmented mast and a variety of antenna bases for use on rental cars, sailboat stern rails, condominium balcony railings or anyplace else I might want to use the radio.

## Splicing the Mast

First, I had difficulty locating 3/8-24 (UNF) hardware to fit most HF mobile antenna bases and the Hustler resonators. Common bolts in hardware stores are 3/8-16 (UNC). I needed a way to fit 3/8-24 threaded studs onto aluminum tubing and join the sections. My solution is to use a piece of 1/2-inch aluminum tubing with a 0.065 inch wall thickness. This leaves an inside diameter of 0.370 inch, five thousandths under 3/8 inch. I bought some 3/8-24 stainless steel bolts, cut off the heads and then ran a 3/8 inch drill a few inches into each end of each tube. The drilling is fairly easy because it only removes a few thousandths of an inch.

The unthreaded ends of the headless bolts slide snugly into the tubes, with about 3/8 inch of threads protruding as a stud. To secure each assembly, I mounted it in a vise and cross drilled (a drill press is helpful, but not necessary) the tube and bolt for #8-32 hardware (see Figure 1) that locks each shaft in place. Some antioxidant paste in the joints prevents corrosion and ensures conductivity. Combine the mast sections by screwing one end of each into a 3/8-24 stainless steel coupling nut.

The aluminum tubing comes in 72 inch lengths, and I discovered that I had exactly enough to make two mast sections for the HF antenna and a piece 19 1/8 inches long for 2 meters. To suit my ICOM 706 that covers HF, 6 and 2 meters, I have an antenna for any HF band (with suitable resonators), a 6 meter antenna (the 54 inch mast alone) and a $\lambda/4$ antenna for 2 meters (the 19 inch tube). Next, I needed antenna bases and mounting options.

## A Mixture of Mounts

A Hustler MBM magnet mount works fine on the trunk lid of a rental car. (I wouldn't try driving at freeway velocities with the HF antenna on this mount, but it's okay for the 2 meter radiator or at low speeds.) I also bought a Radio Shack 21-937 mount, which works well on small horizontal pipe rails such as those found on boats. A small C-clamp or a couple of wood screws can easily attach the RS mount to a wooden deck railing. Attach $\lambda/4$ wire radials to the clamping bolts.

To make a free-standing antenna for a deck, dock or lawn, I cut a circle of thin steel plate about eight inches in diameter and put three bolts through it to attach radials. The ground connection couples capacitively to the steel plate through the mag mount, and the plate must be large enough to accomplish this. For my first test of this arrangement, I used one of my wife's steel cookie sheets with the radials secured to a bolt through the hole in the handle. (Thankfully, the neighbors didn't see me sitting on the deck working Houston on a cookie sheet!) Use monofilament fishing line as

[1]Notes appear on page 17.

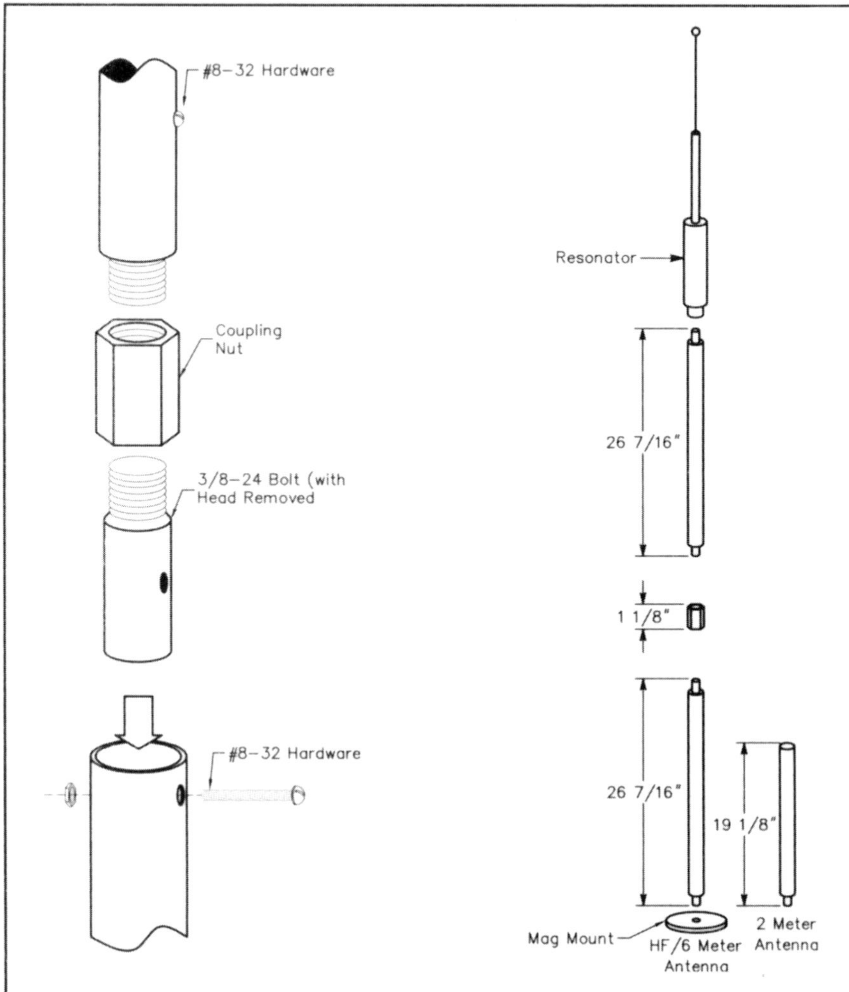


Figure 1—Construction details of N8GGI's portable antenna. Numbers in parentheses are McMaster-Carr catalog numbers (see Note 2).

Materials list:
Tubing—1/2 inch OD, 0.065 inch wall, 6061-T6 aluminum (#89965K54)
Bolts—3/8-24×3 inch long, stainless steel, hex head (#92198AQ362)
Coupling nuts—3/8-24 stainless steel, (#91811A031)

guys in windy situations.

By replacing the untuned car body with λ/4 wire radials, the bandwidths of the resonators seem to be greater than twice what Hustler claims for mobile operation.[1] Using only two radials, I get about 125 kHz on 40 meters and 250 kHz on 20 meters at an SWR of less than 2:1. Antenna resonance changes with environment and the particular mount used. You can regain resonance by adjusting the resonator tip rod according to the manufacturer's recommendations.

## A Suitable Case

My final design task was to create a carrying case that prevents airline baggage "gorillas" from killing the antenna. I use a piece of 4 inch Schedule 40 PVC pipe and two end caps to make a shipping tube. Use glue or screws to secure one end cap. Nylon straps (with plastic catches like those used on fanny packs and life jackets) hold the removable cap. Pop rivet the nylon straps to the pipe and cap. Rivet on a strap handle for convenience. The only items that don't fit into the shipping tube are the magnetic mount and the steel radial plate. A tube large enough to accommodate these pieces would be too heavy and unwieldy.

All the parts for this project come from local fastener distributors and metal dealers. You can also buy them from a commercial supplier, such as McMaster-Carr Supply Company in Chicago.[2]

This antenna obviously won't compete with a full-size antenna, but I've had plenty of good signal reports from the US and Europe while running 100 W. Variations of this theme should allow some pretty stealthy antennas for those in apartments, condos or under antenna restrictions. Run the wire radials through a flower bed or along a deck railing to hide them. You can disassemble the antenna in seconds when the "antenna police" are patrolling.

### Notes

[1]This may seem an improvement, but increased bandwidth for simple antennas usually indicates increased losses. More radials should improve the performance.—*Ed.*

[2]McMaster-Carr Supply Co, PO Box 4355, Chicago, IL 60680-0330; orders tel 630-833-0330, fax 630-834-9427. Their current home page is at **http://www.mcmaster.com/.** They sell to commercial customers and won't ship catalogs to individuals, but they're planning a Web catalog for this fall. Individuals may be able to place orders there.

*Dennis was first licensed in 1961 as K8UKU and got back into hamming in 1984 with his present call. He is a graduate industrial designer and owns a custom exhibit and manufacturing firm in Columbus employing 25 people. His last* QST *article was "Build it Yourself with Plastics,"* QST, *Oct 1987. You can reach Dennis at 133 N Lyndale Dr, Westerville, OH 43081.*

QST

# The Ultimate Portable HF Vertical Antenna

**Phil Salas, AD5X**

AD5X

**The complete unassembled antenna.**

I've had a tremendous response to the portable antenna project published in the July 2002 *QST*. Many folks had problems locating some of the parts, however. It seems that sprinkler-system parts are not as common in other sections of the country as they are in Texas! Also, the original article required that you drill and tap screw holes in brass couplings and the sprinkler-system risers. I wanted to eliminate the screw-hole tapping to simplify the construction.

Over the years, the antenna evolved from the original design through changes that used fiberglass, aluminum tubing, and then brass tubing. The most recent design is longer, lighter and more compact (when disassembled) than the original antenna. It is also easier to build, and easier to the find parts. Band coverage is also increased to include 60 meters, as well as 40 through 10 meters!

The Ultimate Portable Antenna is designed for easy transport. It breaks down into multiple mast sections, a whip section, an air-wound center loading coil section, and a small base support. No piece is longer than about 20 inches, so it will easily fit into most suitcases. The lead photo shows the disassembled antenna, including guys and radials. The fully assembled antenna has a length of almost 16 feet! See Figure 1. That's me, standing next to the antenna set up on my front lawn, in Figure 2.

The key to efficiency for short antennas (such as a loaded vertical less than a quarter wavelength) is the length. The longer the antenna, the greater the radiation resistance and therefore the less impact you have on efficiency due to ground and loading coil losses. This antenna is almost $^1/_4$ $\lambda$ on 20 meters. For operation on 17 through 10 meters, you will shorten the antenna to $^1/_4$ $\lambda$. The length of the complete antenna minimizes the required loading coil for 60, 40 and 30 meters. So let's build it!

AD5X

Figure 1—The complete antenna setup in the author's front yard.

AD5X

Figure 2—Standing next to the antenna to illustrate the height of the tubing sections and coil.

## Gathering the Parts

You can find most of the parts for this antenna at your local hardware store. The loading coil, coil taps, 10 foot telescoping whip, and SO-239 are available from MFJ (**www.mfjenterprises.com**). Be sure to mention this *QST* article when ordering the kit of parts (MFJ-1964-K) for a special price discount. Table 1 is the complete parts list.

A few notes about pipe sizes may be in order for anyone not familiar with plumbing fixtures. When you go to the hardware store to purchase the $^1/_8$ inch NPT nipples and couplers, don't expect to find anything measuring $^1/_8$ inch! The outside diameter of a $^1/_8$ inch pipe is 0.405 inches, or about $^{13}/_{32}$ inch. The standard threads are 27 turns per inch. (The $^1/_8$ inch designation comes from the approximate inside dimension of the pipe, although today you may find pipes with different wall thicknesses and the same outside dimensions. For this project you don't have to worry about the wall thickness of your fittings.) The NPT specification means National Pipe Thread. There are several other thread specifications, but NPT is the most common.

I used $^1/_8$ inch NPT close nipples for the

**Table 1**
**Parts List**

| Qty | Item |
|---|---|
| 1 | 5 inch long × 2.5 inch diameter × 10 TPI air wound coil (MFJ-404-008)* |
| 1 | 10 foot telescoping whip (MFJ-1954)* |
| 1 | SO-239 chassis mount connector (MFJ-610-2005)* |
| 5 | Coil clips (MFJ-605-4001)* |
| 2 | 3 foot pieces of 3/8 inch diameter brass tubing (ACE Hardware) (McMaster-Carr 8950K581 — 6 foot length)** |
| 1 | 3/8 inch diameter wood dowel. (36 inch length at Home Depot—only 3 1/2 inch needed.)*** |
| 1 | 3/4 inch PVC T — All of the PVC fittings are white schedule 40 pipe (Home Depot) |
| 1 | 3/4 inch slip × 1/2 inch female pipe thread PVC adapter (Home Depot) |
| 1 | 3/4 inch slip × 1/2 inch slip PVC adapter (Home Depot) |
| 1 | 3/4 inch slip PVC plug (ACE Hardware) |
| 1 | 1/2 inch NPT (male thread) to 1/8 inch NPT (female thread) brass adapter bushing (ACE Hardware)(McMaster-Carr 50785K64) |
| 8 | 1/8 inch NPT brass couplings (Home Depot)(McMaster-Carr 50785K91) |
| 4 | 0.7 inch long 1/8 inch NPT all-thread nipples (these are also called "close nipples") (Home Depot) (McMaster-Carr 50785K151) |
| 2 | #8 brass wing-nuts (ACE Hardware) |
| 2 | #8-32 × 3/4 inch brass machine screws (ACE Hardware) |
| 2 | #8-32 brass nuts (pack of 6) (Home Depot) |
| 2 | #8 copper-plated steel (or brass) split lock washers (ACE Hardware) |
| 1 | 36 inch length of 1/8 inch diameter brass rod (Home Depot) |
| 1 | 3/8-16 × 1 1/4 inch hex head bolt, zinc plated. Choose the longest you can that is threaded all the way to the head. (Home Depot) |
| 1 | 3/8-16 × 12 inch hex head or carriage bolt, zinc plated. Choose the longest bolt you can find in this size if you can't find a 12 inch bolt (Home Depot) |
| 1 | 3/8-16 coupler, zinc plated (ACE Hardware) |
| 2 | 3/8-16 nuts, zinc plated (pack of 6) (Home Depot) |
| 1 | 3/8 inch lock washer, zinc plated (pack of 10) (Home Depot) |
| 4 | #6 stainless steel 3/8 inch sheet metal screws (ACE Hardware) |
| 3 | #8 solder lugs (Home Depot) |
| 6 | #8 × 1 1/2 inch brass wood screws (pack of 6) (Home Depot) |
| 90 feet | Wire (any gauge, insulated or not, for six 15-foot ground radials.) |
| 1 | Alligator clip and miscellaneous short pieces of connecting wire. |

*The total retail price for all the MFJ parts is $53.80, plus shipping/handling. Mention this article when you order and MFJ will sell a kit containing the telescopic whip, coil, coil clips, and SO-239 connector for $39.95 plus $6 s/h. The kit is MFJ-1964K. You can order additional coil clips for $2.95 each.

**You can use 3/8 inch aluminum tubing if you prefer. Aluminum tubing is about half the cost of brass, but you will either need to drill the brass couplings and aluminum tubing so they can be connected with stainless steel sheet-metal screws or solder them together using aluminum solder and a torch (see **www.solder-it.com**).

***You can use 3/8 inch fiberglass rod instead of the wood dowel if you prefer, but fiberglass rod is more difficult to find. Check out bicycle flags and driveway marker stakes as potential sources of fiberglass rod.

The brass plumbing items are also available from McMaster-Carr, (**www.mcmaster.com**) a mail-order supplier with no minimum order requirement. They stock 6 foot lengths of 3/8 inch brass tubing as well as the close nipples, couplings and brass bushings. The Table lists part numbers to help you look up those items on the Web site.

antenna connectors. These are about 0.7 inches long, and are usually fully threaded over over the entire length. The PVC pipe fittings must be the white, schedule 40 PVC pipe. Do not use the thinner-walled pipe sometimes known as CPVC, which has a cream color or yellow tint. When fittings are designated as "slip," it means pieces are intended to be glued—they just slip together—rather than being threaded.

## Brass Rod Preparation and Assembly

See Figure 3 for the assembly details. First cut three 18 inch sections of the 3/8 inch brass tubes with a hacksaw or tubing cutter and de-burr the tubing. The ends of the couplings that fit over the brass tubes must be reamed out with a 3/8 inch drill bit. Otherwise the couplings won't fit over the tubing. To do this, first screw a 1/8 inch NPT coupling on each end of a 1/8 inch NPT close nipple. Use wrenches to screw these together as tightly as possible. Next, clamp one of the couplings securely and ream out the opposite coupling with a 3/8 inch drill bit. (Use a drill press for this operation if at all possible.) Reverse, and ream out the other coupling. [See Figure 4 for one way to use a woodworker's clamp and a drill press.—*Ed.*] Now unscrew the couplings. One end of the nipples will break loose from one coupling, and the other end will stay tight in the remaining coupling. You'll now have a female and male end that will fit over each end of a section of brass tube, as shown in Figure 5. You will need four pair of these male/female brass connectors: three pair for the brass tubes and one pair for the loading coil assembly. If you'd like, you can solder the nipple/coupling assemblies together. The assembly tends to be very tightly secured even without soldering, however.

Now insert the male/female brass pairs

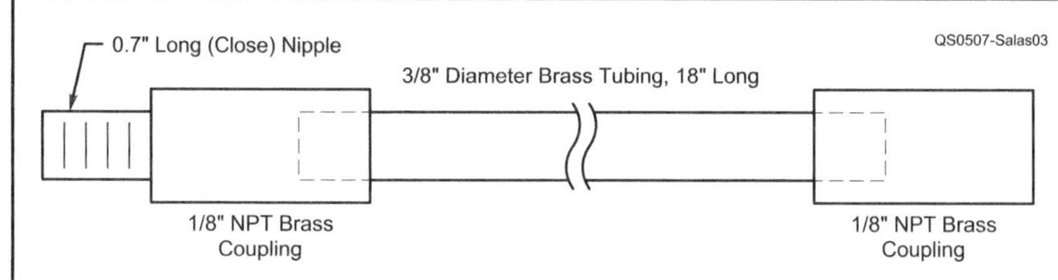


**Figure 3—Assembly of the brass tubing sections.**

**Figure 4—A pipe coupling and nipple secured in a woodworker's clamp ready to drill out the end on a drill press.**

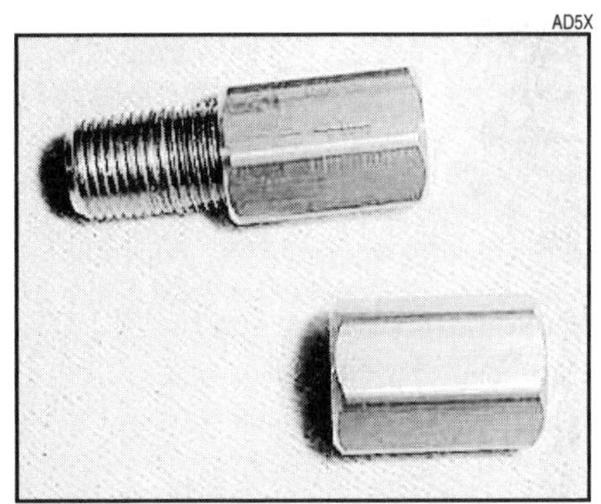


**Figure 5—After the pipe couplings have been drilled out and one coupling removed from the nipple, the pair is ready to be installed on the ends of an 18 inch length of brass tubing.**

just constructed over all three of the 18 inch brass tubes and solder the couplings directly to the tubes. This is easily done with a large soldering iron, or even better, with a torch and silver solder. Solder-It has a nice small butane torch that works well. See **www.solder-it.com**.

## Loading Coil Assembly

Slide $^{1}/_{8}$ inch NPT male/female coupling pairs over both ends of a $^{3}/_{8}$ inch diameter, $3^{1}/_{2}$ inch long wood dowel. You will need to drill a $^{1}/_{8}$ inch diameter hole completely through each of the $^{1}/_{8}$ inch NPT brass couplings and dowel as shown in Figure 6. Next cut two 3 inch lengths of the $^{1}/_{8}$ inch diameter brass rod. Insert one of these 3 inch sections through the holes on one brass coupling. Center the rod so that equal lengths are available on both sides of the coupling, and solder the rod to the coupling with a large soldering iron or torch. (Be careful not to burn the wood dowel with the torch!)

Now position a 3 inch length of the MFJ-404-008 coil such that the 3 inch brass rod just installed pokes through the last two turns on the coil. See Figure 7. Solder the coil turns to the rod. On the opposite end of the coil assembly, insert the remaining 3 inch brass rod through two adjacent turns on this end of the coil, through the brass coupling, and through the coil turns. Solder the coil turns to the brass rod, and then solder the brass rod to the coupling.

Next, indent every other turn on the coil with a small flat-head screwdriver. You should do this on opposite sides of the coil to give you plenty of adjustment capability.

Finally, on the end of the coil with the brass nipple (male end), solder a 6 inch piece of insulated wire terminated with an alligator clip. For extended outdoor use, you may wish to treat the wood dowel with varnish.

## Top Whip

The MFJ-1954 10 foot telescopic whip comes with a standard $^{3}/_{8}$-24 mounting thread. While the whip mounting stud is not a correct match for the $^{1}/_{8}$ inch NPT coupling (27 turns per inch), I just thread the screw directly into the $^{1}/_{8}$ inch NPT coupling. Because the $^{3}/_{8}$ bolt is slightly smaller than the $^{1}/_{8}$ inch pipe outside diameter, and the pipe coupling is tapered inside, they will go together. [See the "Alternative Construction Ideas" sidebar for a way to match the threads, if you are uncomfortable with the mismatch.—*Ed.*]

## Base Assembly

For the base spike, I've used a $^{3}/_{8}$-16 × 12 inch zinc-plated hex-head bolt. Only $1^{1}/_{2}$ inches of the bolt is threaded, and so I used the long smooth end of the bolt to go into the ground after cutting off the hex head. A damp cloth easily cleans the bolt after use. I also used a $^{3}/_{8}$-16 × $1^{1}/_{4}$ inch zinc-plated hex-head bolt at the base of the PVC assembly, and a $^{3}/_{8}$-16 zinc-plated coupler to attach the $1^{1}/_{4}$ inch bolt to the 12 inch bolt, as shown in Figure 8. This way you can leave the long bolt off if you want to bolt the base assembly directly to a metal plate or trailer mount, or screw on the long bolt for ground mounting.

Referring to Figure 8, drill a $^{3}/_{8}$ inch diameter hole into the $^{3}/_{4}$ inch PVC plug used for the base support $1^{1}/_{4}$ inch bolt. Cut off about half of the length of the $^{3}/_{4}$ inch PVC plug to leave plenty of room inside the T for wiring. Solder a ground wire to the head of the $^{3}/_{8}$-16 × $1^{1}/_{4}$ inch bolt as shown, or use a $^{3}/_{8}$ inch solder lug. Insert the threaded end of bolt into the plug, and secure with a $^{3}/_{8}$-16 nut and a lock washer. If you wish, you can glue the plug in place with PVC pipe cement instead of using the #6 stainless steel sheet metal screws shown. The screws make changing the support assembly easy in case you should ever want to, though. You might even consider using screws to attach the other two adapters into the T.

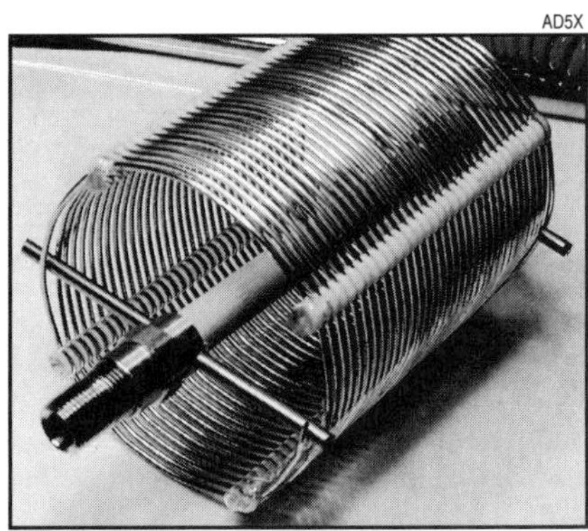


**Figure 7—The loading coil assembly ready to be soldered.**

To prepare the 12 inch bolt, cut off the hex head and round this end with a file. Screw a $^{3}/_{8}$-16 nut onto the threaded end, then add a lock washer and screw the $^{3}/_{8}$-16 coupler onto the threads so the bolt is about halfway through the coupler. Tighten the nut against the coupler (with a lock washer between the nut and coupler). This 12 inch bolt assembly can now be easily screwed onto the $1^{1}/_{4}$ inch bolt on the base assembly for ground mounting.

Now place the SO-239 over the $^{1}/_{2}$ inch hole in the $^{3}/_{4}$ to $^{1}/_{2}$ inch slip adapter and mark the location for the two #6-32 × $^{3}/_{8}$ inch long stainless steel sheet metal screws that will hold it in place. (The adapter I used had a hex head on the outside lip, and by turning the SO-239 so diagonal mounting holes are over opposite points in the hex head, there should be enough material to hold the screws easily.) You'll see that these holes will be right in the center of the PVC lip. Carefully drill two $^{5}/_{64}$ inch holes at these points. We will mount the SO-239 on the adapter later.

Place the $^{3}/_{4}$ inch PVC plug/spike assembly in the T and drill two $^{5}/_{64}$ inch diameter holes through the T and plug. [See sidebar Photo A for how I clamped the T in a woodworker's clamp to drill the holes. Try to drill straight through both sides in one pass to make alignment easier.—*Ed.*] Remove this

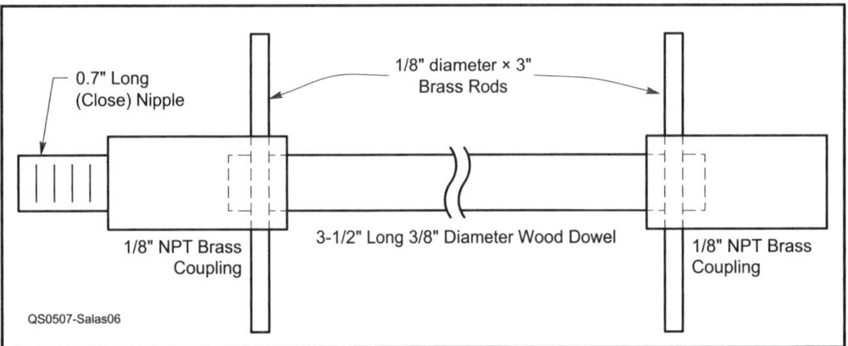


**Figure 6—How brass couplings are installed on a length of $^{3}/_{8}$ inch wood dowel and $^{1}/_{8}$ inch brass rod in order to mount the coil.**

## Alternative Construction Ideas

As I began editing this article for publication in *QST*, I decided to build the antenna. After visiting three local Home Depot stores and most of the True Value and ACE Hardware stores in the area, I realized that builders may have difficulty finding some of the materials. For example, none of the stores I visited had any stock of brass or aluminum tubing in the size I wanted.

I contacted Phil Salas with a few questions about this article and the materials he specified. Phil informed me that since writing the article he had continued to try new ideas for building the antenna, and he found several alternative materials. His first suggestion was to use 1/2 inch aluminum tubing, with a wall thickness of 0.058 inch. He tapped the inside of that tubing for 1/8 inch pipe threads, and just turned the brass nipples into the pipe. This saves the step of drilling out the couplings to slip the 3/8 inch brass tubing inside. I was able to find some 1/2 inch aluminum tubing with a 0.050 inch wall thickness but not the thicker wall. Unfortunately, a 1/8 inch NPT nipple slips right inside that tubing, so I could not cut threads into the tubing. Perhaps 7/16 inch tubing with a 0.050 inch wall thickness would work, but I did not find any tubing in that size.

Phil also suggested McMaster-Carr (**www.mcmaster.com**) or similar mail-order/Internet stores as a good way to obtain the tubing and other plumbing parts. Their prices compare favorably with what I found in Connecticut, and with no minimum order they may save you some shopping time in a variety of hardware stores. Of course many of us probably enjoy browsing the aisles of our local hardware stores. You never know what new tool you may need or what other antenna projects may come to mind as you peruse the available materials. Phil has included McMaster-Carr part numbers for several of the pieces in Table 1.

I found a K & S Engineering display of hobby tubing at my local Do It Best hardware store. They ordered the 3 foot lengths I needed for a reasonable price. After building an antenna with that tubing, however, I realized that the 0.014 inch wall thickness was too thin to support the antenna without guy lines. Phil suggested that I could use that tubing by inserting a 1/4 inch dowel or 1/4 inch fiberglass rod, such as from a bicycle flag to stiffen the brass tubing. I ordered a 6 foot length of the brass tubing with a 0.032 inch wall from McMaster-Carr, as listed in Table 1. That proved to be a near ideal tradeoff between weight and strength.

I also discovered a much wider array of PVC pipe fittings than I had imagined. Many of the pieces appear to have similar specifications at first glance.
For example, I found 3/4 inch × 1/2 inch adapters with male pipe threads on the 3/4 inch end and either male or female threads on the 1/2 inch end as well as one with both ends designed for slip fitting of the PVC pipe into the adapter.

There are many different ways to arrive at the end result you need. For example, if you can't find a 1/2 inch × 1/8 inch brass bushing for the top of the T, a 1/4 inch × 1/8 inch bushing will turn into a 1/2 inch × 1/4 inch bushing to achieve the same results.

All of my searching for parts shows that there may be a variety of materials that you could use to build a similar antenna. What you find in your area may not match what the author or I found. With a bit of creative shopping you will be able to find the parts you need to build the antenna.

The author suggests that you use #6 × 3/8 inch sheet metal screws to hold the pipe plug with the bolt into the bottom of the T. This will make it easier to disassemble that section if you have to fix a wiring problem inside the T. I decided that was a good approach for the fittings on the other two parts of the T as well. Photo A shows the T clamped in my woodworkers clamp, ready to drill the screw body holes.

WR1B

**Photo A — The 3/4 inch PVC T is clamped in a woodworker's clamp, ready to drill the screw-body holes with a 9/64 inch drill bit. I had first drilled 5/64 inch pilot holes with the adapters and bottom plug installed in the T. Try to center the hole on the side of the T, and drill all the way through both sides. Then remove the adapters and plug, and drill the screw-body holes in the T.**

Phil has found that threading the 3/8-24 stud on the bottom of the telescoping whip into the 1/8 inch NPT coupling is adequate. The perfectionist in me doesn't like the idea of mismatched threads. I decided to make a slightly different adapter than Phil suggests in his article. I used a 1/8 inch pipe coupling and nipple on the bottom of a 1 inch length of the brass tubing. (I drilled out the brass coupling with a 3/8 inch drill, the same as is done for the antenna tubing sections.) For the top section of my adapter I used a 3/8-24 coupling nut, which is about 7/8 inch long. I drilled out the threads on about half of that coupling, too, and pushed it onto the brass tubing. (I soldered these, and all other joints on my antenna.) Now all the threads mate properly. The 1/8 inch NPT to 3/8-24 adapter allows me to attach the whip to any antenna section, as desired.—*Larry Wolfgang, WR1B*

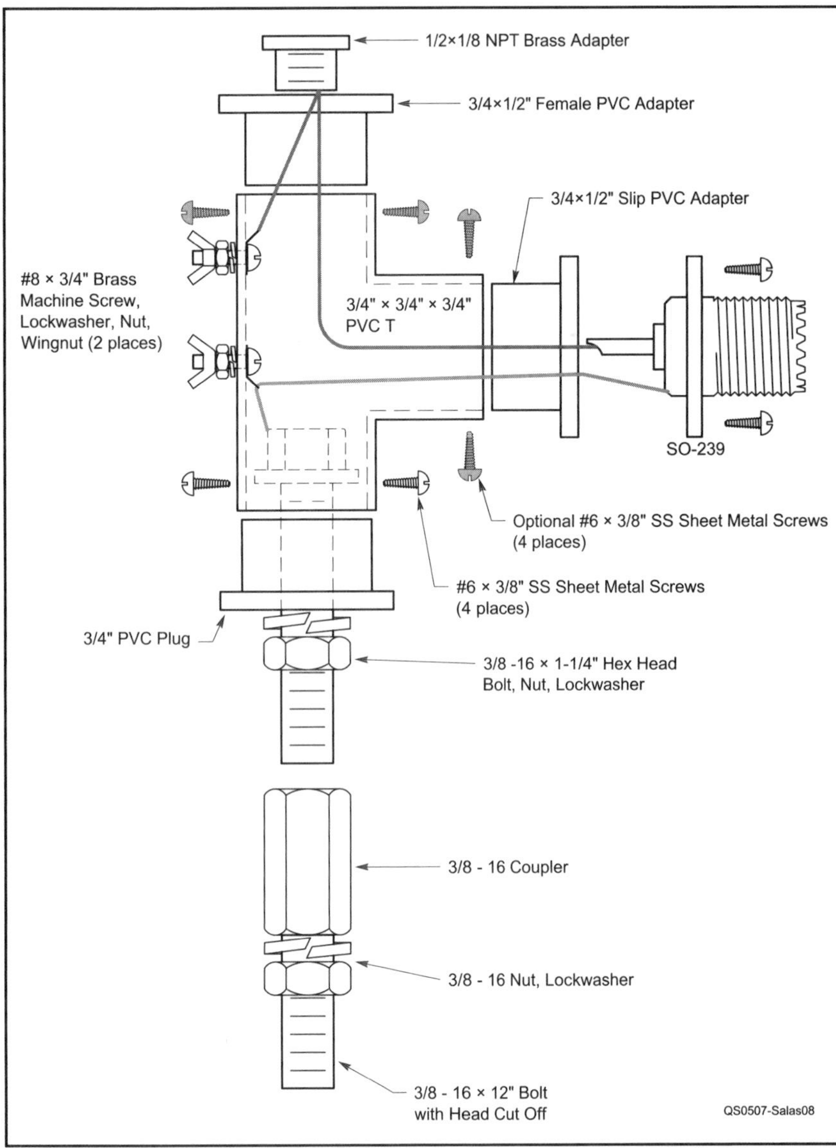


**Figure 8—The antenna base and coaxial cable connector details.**

assembly from the T and drill out these 5/64 inch holes in the T to 9/64 inch. Also drill out two holes in the SO-239 connector to 9/64 inch, since the holes are not large enough to pass the #6 × 3/8 inch sheet metal screws.

Use an 11/64 inch drill bit to drill mounting holes for the #8-32 brass machine screws. I drilled the mounting holes close to the "top" and "bottom" of the T. Alternatively, if you align the bit with the inside edges of the side of the T and drill through the back of the T, you will be able to fit a screwdriver into the T to tighten this hardware.

Next we'll prepare the antenna interface at the top of the base. First, cut off part of the 3/4 inch slip × 1/2 inch female pipe thread PVC adapter so as to leave additional room in the T for wiring. Solder a piece of 14 gauge copper house wire directly to the inside lip of the 1/2 × 1/8 inch NPT brass adapter. You'll need a large soldering iron or a torch, since the brass adapter mass is pretty large. Screw this adapter tightly into the 3/4 × 1/2 inch PVC adapter.

Now solder a wire to the center conductor of the SO-239 connector as shown. This wire should be soldered to the wire stub on the 1/2 × 1/8 inch NPT brass adapter at the antenna interface, and then to the upper wing-nut assembly as shown. (Alternatively, you could use #8 solder lugs for these connections, and put the top #8-32 brass machine screw through the hole in the solder lug.) The 3/4 × ½ inch PVC adapter can now be glued into place using PVC pipe glue. Solder a short piece of copper braid (from a piece of RG-58 cable) from the SO-239 ground (solder directly to the SO-239 body) to the brass ground screw, and finally to the wire soldered to the head of the 1 1/4 inch bolt. (#8 solder lugs are a good alternative.) You can now complete the assembly of the base by inserting the PVC plug and 1 1/4 inch bolt assembly into the T and installing the #6 stainless steel sheet metal screws as shown in Figure 8. Incidentally, the upper wing-nut assembly is used in case you need to add capacitive or inductive base matching should you want to improve the SWR on the lower bands. See Figure 9.

## Ground Radial Network

The radial network is made up of six 15 foot radials, using 22 gauge insulated wire, though any gauge wire, insulated or not, can be used. I've found it best to make up three pairs of two wires each attached to a #8 spade lug on one end of each pair. This minimizes the hassle of deploying, and later rolling up, the radials. The three #8 lugs will attach to the ground screw on the base assembly. When the wires are rolled up, you should hold them together with twist-wraps. Solder a 1 1/2 inch brass wood screw on the outer end of each radial. You can simply push these screws into the ground to hold the radials in place. Put a blob of hot glue on each wire/screw soldered interface to give it a little strain relief.

## Guying

This antenna is self-supporting in a low breeze. In many cases, however, it will be necessary to guy the antenna because of its 16 foot length. For effective guying, I attached 9 foot lengths of nylon cord (3 pieces) just above the base of the 10 foot MFJ telescoping whip. Use a tie-wrap and close it just enough so that it won't slide over the base of the MFJ whip. Cut the 9 foot sections of nylon cord and heat the ends with a match to fuse the nylon so it won't unravel. Tie one end of a 9 foot section of nylon cord around each tie-wrap and secure with hot glue or epoxy. For the ground stakes, you can use the extra piece of brass tubing. (You only used 4 1/2 feet of the 6 foot length.) Cut the remaining 18 inch piece of tubing into three 6 inch sections. Attach the end of the nylon cord without a tie-wrap to one end of each tube with hot glue. Also plug the open ends of the three tubes with hot glue. (You could also use some long spikes or bolts for stakes, if you are concerned that you won't be able to drive the brass tubing into hard ground.) For storage, wrap the nylon cord around each brass stake and hold it in place with masking tape. Figure 10 shows the details. See Figure 11 for a photo of the guys attached to the telescoping MFJ whip. When bolted to a trailer mount or plate, the antenna should really

Figure 9—The completed antenna base and mounting spike.

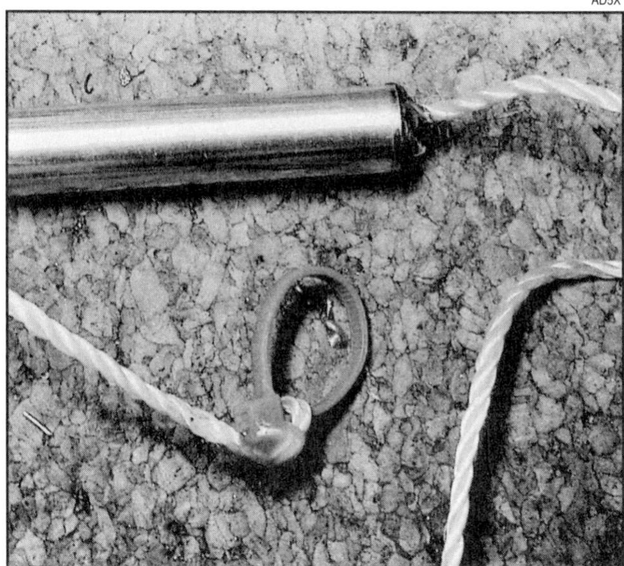

Figure 10—A set of 3 guys was made by using a wire tie (closed to slip over the top of the telescoping whip but not over the base of the whip). Hot glue was used to attach a nylon line to each wire tie. Hot glue was also used to attach the guy line to a left-over length of brass tubing.

not need guying unless the wind is strong.

## Antenna Assembly

To assemble the antenna, first screw the three brass-rod sections together, and then screw these into the top of the base assembly. Push the entire assembly firmly into the ground, keeping it as vertical as possible. Next, screw the loading coil and telescoping whip assemblies together. Slip the three guy tie-wrap/nylon cord assemblies over the whip and extend the telescoping whip. Screw this entire top assembly into the female end of the top brass tube. You only need to turn the brass fittings finger tight. Finally, push the guy rods in the ground and extend the six radials. Attach the common ends to the ground screw on the base assembly. Add coax, and you are ready to tune your antenna.

## Antenna Tuning

Begin with 60 meters to determine the tap points on the coil. If your rig cannot tolerate a 2:1 SWR, you may need to add a 330 pF, 300 V, silver mica capacitor across the two wing-nut assemblies. See Figure 12 for a close-up view. This capacitor is fine for both 60 and 40 meters. The SWR on 30 meters will be closer to 1.7:1. If you need to improve this, use a 220 pF capacitor. You should not need any capacitors for 20 through 10 meters. Both my IC-706MKIIG and SG-2020 work fine into a 2:1 SWR.

To begin, use an antenna analyzer set to 5340 kHz to locate the coil tap point that gives the best SWR. Mark this tap point. Repeat the procedure at 5380 kHz. Move to 40 meters and repeat, again selecting two taps on 40 meters to give you the band coverage you desire. Repeat again for 30 meters—only a single tap is required for this band. For 20 meters, you will find that only the top turn of the coil is necessary for resonance. The antenna is almost a quarter wavelength long on 20 meters.

For 17, 15, 12 and 10 meters, you will need to short out the loading coil and remove sections of the brass tube—then adjust the whip for resonance. On these bands, the antenna will be $^1/_4$ $\lambda$ long. Remove two brass sections for 17 and 15 meters, and all three sections for 12 and 10 meters. Use a permanent black marker pen to indicate the correct band positions on the telescoping whip.

Finally, attach the MFJ-605-4001 coil

Figure 11—A close-up of the guy lines attached to the telescoping whip. The coil and coil clips detail can also be seen.

Figure 12—The antenna base, with the ground radials and a matching capacitor attached to the brass screws on the PVC T.

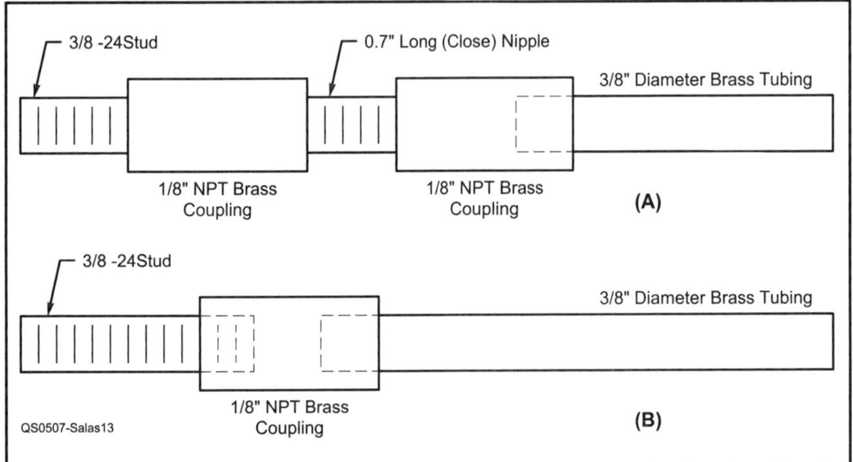


**Figure 13—(A) shows how a 1/8 inch NPT brass coupling can be used to adapt the pipe threads to a 3/8-24 standard antenna mount. (B) shows an alternative attachment, if one only wants to attach the antenna to a standard mount, and not use the PVC base to mount the antenna on the ground.**

clips to the coil tap points determined above. You may wish to solder the clips in place to make the entire assembly a little more robust. From this point forward, you can just go back to these tap points, or re-adjust the top whip as necessary, and not have to worry about making SWR measurements.

## Mounting Options

You can easily make a 3/8-24 threaded interface so that the antenna can be mounted on a standard 3/8-24 antenna mount. This would be useful for those with a standard ball mount on their car who want to use this extended-length antenna when stopped. As mentioned earlier, the 1/8 inch NPT thread is 27 turns per inch, and it is slightly larger than 3/8 inch diameter, with a slight taper. While the 3/8-24 standard stud will screw into an 1/8 inch NPT thread coupling, the 1/8 inch NPT nipple will not screw into a 3/8-24 threaded coupling. Therefore, an adapter is necessary if you want to mount this antenna to a standard 3/8-24 antenna mount.

One way to make an adapter is to purchase a 3/8-24 bolt and screw it tightly into a 1/8 inch NPT coupling. Cut off the head of the 3/8-24 bolt with a hacksaw and file carefully so that you don't damage the threads. Running a 3/8-24 die over the threads will clean up any damage you may have done. You can now either screw this assembly onto the 1/8 inch NPT nipple on the bottom brass tube section, as shown in Figure 13A, or screw the 3/8-24 bolt directly into the bottom 1/8 inch NPT coupling as shown in Figure 13B (if you'll never need the 1/8 inch NPT interface on the bottom antenna section).

## Conclusion

Because of the interest in my original portable antenna, I've evolved that design into an antenna that is longer, lighter, more compact and easier to fabricate, and gives you more mounting options. You can also experiment with the antenna length. For example, you can remove a section or two, use more or fewer sections, decrease or increase section lengths, or place the loading coil in different positions. With the loading coil described in this article, you have quite a bit of latitude in the antenna length for a given band. For best efficiency though, try to keep the antenna as long as possible and the coil as high as possible.

Don't hesitate to make changes based on hardware availability. Try aluminum or copper tubing, or even wire wrapped 3/8 inch fiberglass or wood dowel. It's fun to design antennas "on the fly" while standing in the plumbing section of your hardware store. This makes for interesting discussions with the clerks, however!

Finally, I want to express my appreciation to Martin F. Jue, K5FLU, and Richard Stubbs, KC5NSZ, of MFJ, for working with me to provide a reasonably priced kit of parts to make this both an affordable and fun project to build.

*Phil Salas, AD5X, is an ARRL Life Member. He's been licensed for 41 years, and enjoys HF operating (mostly CW). Phil's wife Debbie (N5UPT) and daughter Stephanie (AC5NF) are obviously very understanding of this hobby! Phil holds a BSEE from Virginia Tech, and an MSEE from Southern Methodist University, and is now fully retired after 33 years in the telecommunications industry. Phil can be reached at 1517 Creekside Dr, Richardson, TX 75081-2913 or at* **ad5x@arrl.net** *if you have any questions or comments.*

By Ron Herring, W7HD

# A Small, Portable Dipole for Field Use

The title says it all—a practical, transportable antenna for Field Day and a valuable addition to your emergency kit.

This antenna came about because I wanted something small and portable that could be used on any band and would perform just like a dipole. Since I didn't want to have to find two trees the right distance apart (difficult to do in arid Arizona) for stringing up a regular dipole, something of a more "stand-alone" nature was needed. Looking through my accumulation of antennas, I discovered that I had a pair of Hamstick mobile antennas for 20 and 40 meters.[1] Needing something for my brand-new PSK20 rig for Field Day, I decided to try building something that would both fit the bed of my pickup truck and be quickly assembled and tuned at a site.

I hit on the idea of using two of these antennas to build a portable dipole. Since a single Hamstick antenna was designed as a mobile antenna and uses the vehicle as a counterpoise, a pair of these appeared ideal for my purpose. I made a quick trip to the local hardware store to pick up some nuts and bolts; a piece of ¾″ by 4″ hardwood (I used oak); a good quality wooden broomstick and some angle-iron with pre-drilled ³/₈″ holes. I then proceeded to drill, screw, tape, assemble and make it work. The total cost

[1]Lakeview Company, 3620-9A Whitehall Rd, Anderson, SC 29626; 864-226-6990; **www.hamstick.com/**.

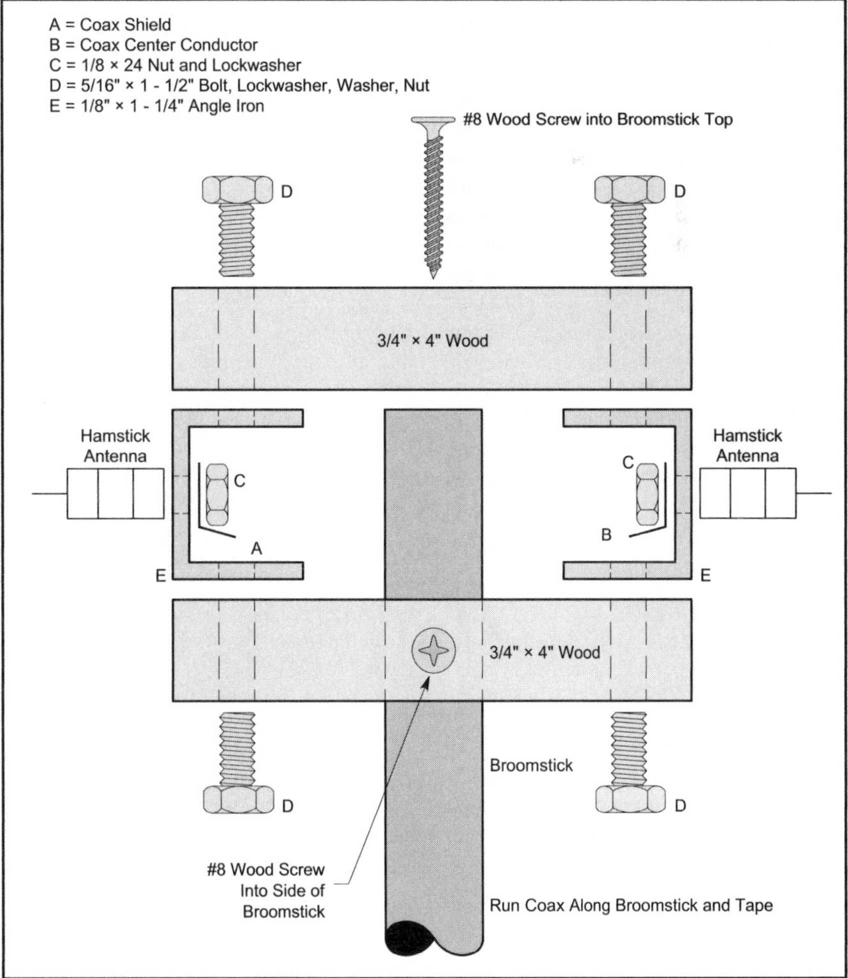


**Figure 1—Assembly details for the portable dipole.**

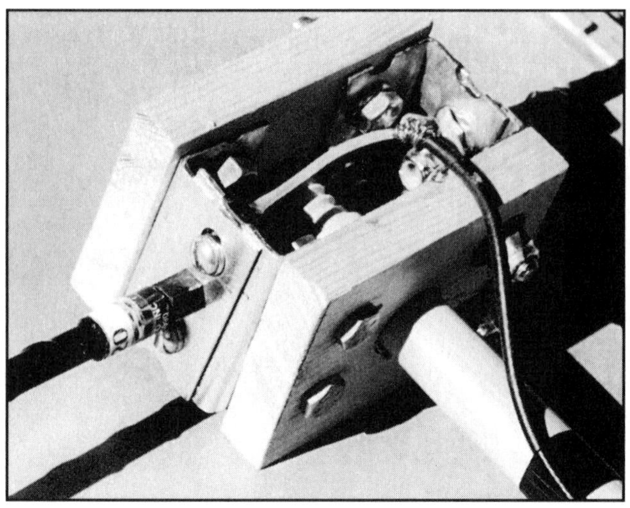

**Figure 2—The completed dipole center support showing the broomstick mast, the antenna mounts and the connected transmission line.**

**Table 1**
**Portable Dipole Parts list**

8—$^5/_{16}$ × 1$^1/_2$″ bolts with lock washers, flat washers and nuts (nylon-type insert hold best) for mounting the angle iron "U" to the wood pieces.
4—$^5/_{16}$ × 1″ bolts with lock washers, flat washers and nuts for assembling the angle iron "U" pieces.
4— $^1/_8$″ angle iron cut to 3″ lengths (cut so holes line up when mating).
2— $^3/_8$″ × 24 nuts with lock washers and flat washers for the antenna mounts.
2— ¾″ × 4″ piece hardwood about 5″ long.
2— $^1/_8$″ × 1$^1/_2$″ wood screws. (I used decking screws).
2— Mobile antennas that use $^3/_8$″× 24 standard thread mounts (I used Hamsticks).
1—Coaxial cable (I used RG-58/U) stripped and tinned to allow connections about 5″ apart.

of materials, including $25 for each of the two Hamsticks, was about $90. Although the Hamstick was available, any suitable shortened (helically loaded) vehicular antenna can be used.

Using nothing more exotic than simple hand tools, a tape measure, power drill, wrenches and screwdrivers, the whole thing came together in about 3 hours. The best part was that it worked exactly as I had planned.

Some tips when you do your own assembly:

• Be sure to tune both antennas on the vehicle before mounting to the assembly.

• For safety, the radiating elements should be out of reach.

• Put a piece of tape on the Hamsticks, marked with the exact length of the "stingers" (the tuning rods) for ease of assembly at the site.

• Treat the wood support with water-seal, lacquer or marine varnish prior to assembly, to prevent deterioration. Just make sure that whatever you use for a coating is non-conductive at RF frequencies.

This antenna will even work on a balcony or supported by a couple of tree branches. My plan is to simply use bungee cords to attach it to the side of the camper at a Field Day site. Since it's a directional antenna, that mounting technique makes it easy to turn. For testing, I simply used bungee cords to attach the antenna to the side of my pickup truck. Table 1 lists the parts necessary to build your own version.

Figure 1 shows how the parts fit together. Figure 2 shows the completed mount. The assembly sequence I used was as follows:

1. Bolt two pieces of angle-iron together to form a "U," making sure that the hole for the antenna is properly aligned. Repeat for the other half.

2. Using the angle iron as a guide, drill two holes in each piece of ¾″ × 4″ wood support to allow the bolts to pass through. Repeat this on both pieces of wood for each side. Make sure that the gap between the angle-iron pieces is more than an inch, since the broomstick has to pass through the gap.

3. Drill a 1″ hole in the bottom piece of wood for the broomstick to pass through. Additionally, drill a $^1/_8$″ hole in the top piece of wood for a wood screw to secure the top of the broomstick. Drill a $^1/_8$″ hole in the side of the top piece of wood for a second wood screw to anchor the broomstick so it doesn't turn in the mount.

4. Assemble as shown in the figure. Mount the angle-iron "U" pieces to the *inside* of each piece of wood. Be sure to attach the coaxial cable to the metal pieces––I just anchored the wire underneath the lock washers.

5. Stick the broomstick through the bottom hole and put the wood screws in place. (Drill a $^1/_{16}$″ pilot hole in the broomstick before anchoring, so it won't splinter.)

6. Tape the coax to the side of the broomstick every 18″, leaving the coax free for approximately the bottom foot of the broomstick.

Okay... it's time to test! Place the antenna in the clear and attach your antenna analyzer or transceiver and SWR meter. Using a low power setting, check the bandwidth of the antenna. It should be about the same as when it was mounted on the mobile mount, perhaps slightly greater. Trim both sides for minimum SWR. Then check the SWR again using full power. Watch for arcing! If arcing does occur, your spacing is too close.

Put a label on the Hamstick giving the length of the "stinger" for the desired operating frequency. You may wish to do this for several favorite operating frequencies. This will save a lot of set-up time at your destination.

*Ron Herring, W7HD, has been licensed since 1967 and worked in engineering for the Heath Company in 1968-69 (SB-103, SB-303, MWW-18). While there, he took a two-year course in computer design, which inspired his future. Working for the Kellogg Co (later Michigan Bell) and then Pacific Northwest Bell (Oregon) he also played a role in the development of the RadioShack Model 100 laptop computer. Ron has taught computer classes at Portland State University, as well as a private school. Currently living in Arizona, he works for the Pima County Sheriff's Department in Tucson, where he is a Network Manager. Ron can be reached at 10270 W Mars Rd, Tucson, AZ 85743;* **w7hd@arrl.net**.

By Phil Salas, AD5X

# A Simple and Portable HF Vertical Travel Antenna

## How to build a portable, efficient antenna without bursting your budget.

With all the small HF rigs available today, the ability to take an HF station on business trips, vacations, family visits, camping and other activities is becoming very easy. Often, the limiting factor is an effective portable antenna to go with the radio. As I'm sure most of you know, the bigger the antenna the better the performance. I usually prefer a full-size dipole. Depending on your location, a dipole may be inconvenient because it needs some form of support. I decided to look into building an inexpensive, portable, vertical antenna for use when a dipole is not practical.

My goal for this antenna was for it to be as long as possible so as to maximize its radiation resistance and hence, efficiency; yet it must also be easily packed in a small suitcase. In addition, it needed to be multiband, covering 40 through 10 meters. The parts needed had to be readily available and, of course, it shouldn't break the bank! Finally, I wanted the antenna to provide low (nearly 1:1) SWR so no antenna tuner was needed.

The resulting antenna described here breaks down into three 2-foot mast sections, a small center-loading coil (air-wound for efficiency), a short telescoping whip and a small base support. When the antenna is put together (a matter of minutes), it has a total height of about 12 feet.

Read through all the directions first to become familiar with the project, but don't be intimidated by all the assembly directions. This antenna is easier to build than it is to explain how to build it. No more than about two hours should be necessary for its construction.

### Gathering the Parts

Except for the loading coil, all parts are available from either your local hardware store or RadioShack. I obtained the loading coil from Surplus Sales of Nebraska (**www.surplussales.com**). The coil (Miniductor 4027) is 2 inches in diameter by 10 inches long, with 10 turns per inch of 16-gauge wire. The cost is $15 for one of these coils. As this price, it is not worth trying to build your own coil, and you'll have enough of the coil left over for other projects (maybe a second antenna for a friend?). The complete parts list is shown in Table 1.

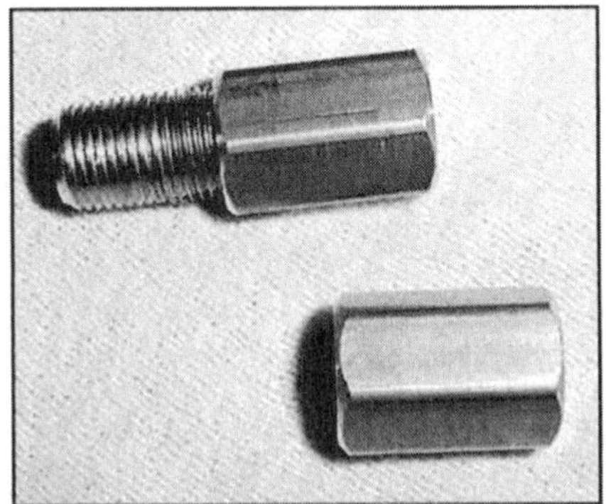
Figure 1—Nipple (top) and coupling (bottom).

### Riser Preparation and Assembly

First, screw each of the three 0.7-inch 1/8-NPT nipples into three separate 1/8-NPT couplings. Screw these in as tight as you can. I used pliers to screw the coupling on

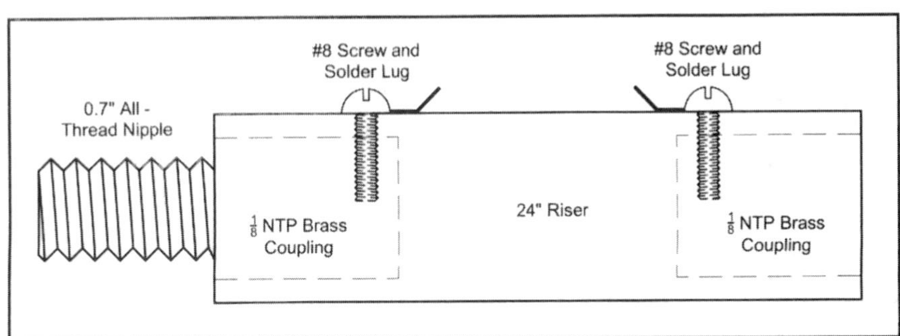

Figure 2—Middle section assembly.

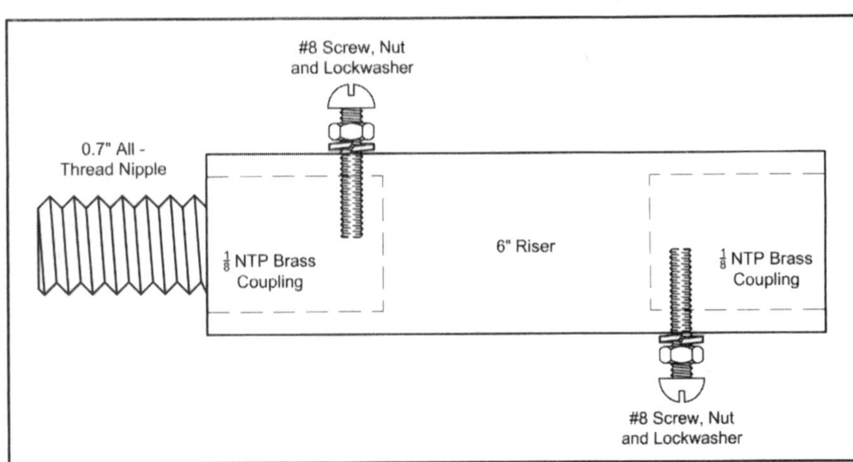

Figure 3—Coil section assembly.

**Table 1**
**Parts List**

| *Qty* | *Description* |
|---|---|
| 1 | Coil, Miniductor 4027, Surplus Sales of Nebraska |
| 3 | 24-inch sprinkler system risers[1] |
| 1 | 6-inch sprinkler system riser |
| 1 | ¾-inch PVC-T |
| 1 | ¾-inch to ½-inch PVC adapter |
| 2 | ¾-inch PVC threaded plugs |
| 7 | ⅛-NPT brass couplings |
| 3 | ⅛-NPT 0.7-inch all-thread nipple "NIPPLE ⅛-inch × CLOSE" |
| 1 | ⅛-NPT 1-inch nipple |
| 2 | #8 × 1¼-inch brass screws |
| 6 | #8 × ½-inch brass screws |
| 2 | #8 brass nuts |
| 8 | #8 copper-plated steel split lock washers |
| 8 | #8 brass flat washers |
| 1 | ⅜ × 12-inch threaded brass rod |
| 2 | ⅜-inch brass nuts |
| 2 | ⅜-inch brass flat washers |
| 2 | ⅜-inch copper-plated steel split lock washer |
| 6 | #6 stainless steel ⅜-inch sheet metal screws |
| 1 | 72-inch telescoping antenna (RS 270-1408) |
| 1 | Chassis-mount SO-239 connector (RS 278-201) |
| 3 | Banana jacks, 2 red and 1 black (RS 274-661) |
| 10[1] | #14 solid copper house wire, insulation removed |
| 10 | ¼-inch solder lugs |

Note: RS is RadioShack.
[1]The adjustable center-loading coil will be a little over 6 feet above the ground. If this is too high, you may wish to change one or more of the 24-inch risers to 18-inch risers, or place the coil assembly between the second and third riser.

each end of a nipple as tight as I could. Then I unscrewed the couplings. One end will break loose right away, and the other will stay tight in the remaining coupling. Next, heat up each nipple/coupling assembly with a large soldering iron and carefully run a bead of solder around the nipple/coupling interface. See Figure 1.

Now insert the couplings *without* the nipples into one end of each of the four sprinkler system risers (one 6-inch and three 24-inch risers). I found this to be a very tight fit on some risers, so you may need to gently tap these in place so they are flush with the ends of the risers.

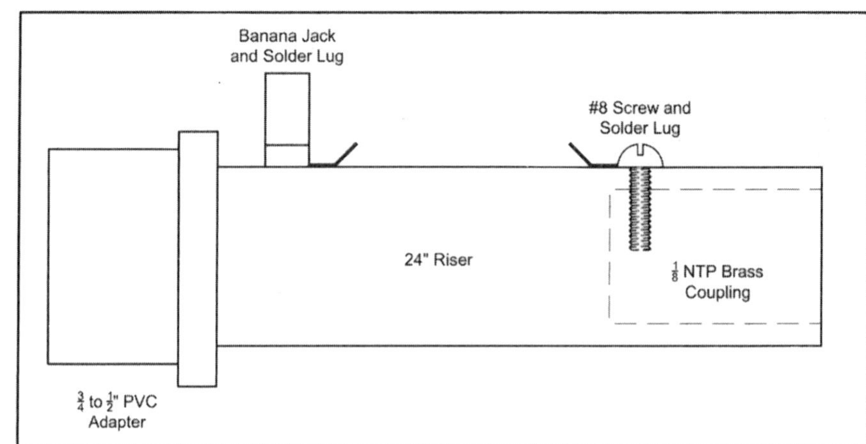

**Figure 4—Bottom section assembly.**

Next insert the remaining three ⅛-NPT couplings (with nipples) into the opposite ends of two of the 24-inch risers, and the 6-inch riser. See Figures 2 and 3. Remember, one of the 24-inch risers does not have a nipple/coupling installed. Again, these nipple/coupling assemblies may need to be tapped into place. To do this without damaging the nipple, insert the coupling-end of the nipple/coupling assembly into the riser as best you can. Then place the nipple-end on a piece of wood, and gently tap the opposite side of the riser with a hammer until the coupling is fully seated in the riser. Make sure that the nipples extend out of the risers.

Next drill and tap a #8 threaded hole through each end of the risers into the inserted ⅛-NPT brass couplings. Each of the inserted ⅛-NPT couplings is 0.7-inches long. Measure back from the end of each riser 0.5-inch and drill a hole (#29 drill, 0.135-inch diameter) through the riser and one side of the brass coupling. The holes should be positioned on the same side of the long (24-inch) risers, and on opposite sides of the short (6-inch) riser. Tap the holes with a #8 tap. Insert the ½-inch long brass screws with lock washers and flat washers into the tapped holes in the 24-inch risers.

Now, insert each of the two #8 1¼-inch brass screws through a #8 nut, lock washer and flat washer. Screw them into the tapped holes on the 6-inch riser. Leave most of the screw protruding out from the riser. Tighten the nut to secure everything in place. These screws will be used for the coil support.

Remember that one of the 24-inch risers did not have a nipple/coupling assembly installed. This riser is the bottom antenna section. To prepare this section, drill a ⅛-inch hole in the riser just above the open threaded end of the riser. Into this hole, mount a red banana jack with its associated solder lug mounted on the outside of the riser. After you get everything tightened, you might want to drip a little epoxy on the mounting screw inside the riser to keep the banana jack from coming loose. Next, screw this end of the riser tightly into the ¾ × ½-inch PVC bushing. See Figure 4.

It is finally time to install the antenna wires. Strip the insulation off of three 30-inch pieces of #14 solid copper house wire. Solder one end of each of the three wires to a #8 solder lug. Attach each of these solder lugs to the brass screw on one end of each of the 24-inch risers. Now wrap two turns of the wire around each of the 24-inch PVC risers and determine where the wire should be cut and another solder lug attached to connect the wire under the far-end brass screw on each riser. The bot-

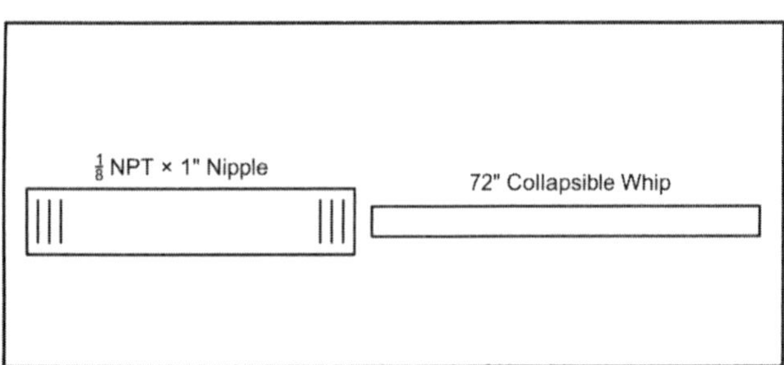

**Figure 5—Collapsible whip assembly.**

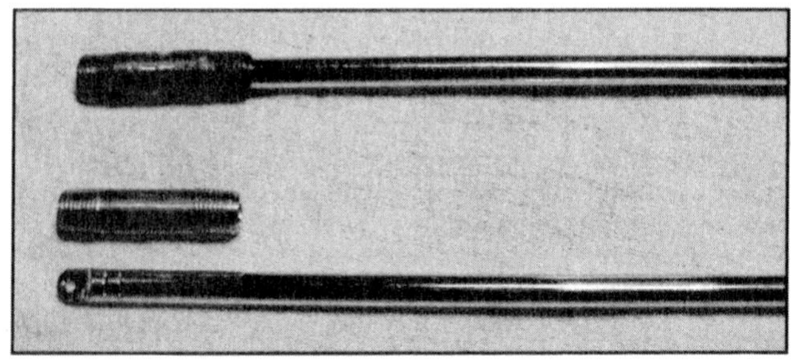
**Figure 6—Whip and coupling.**

Figure 7—Coil assembly.

tom section riser should already have a solder lug attached under the banana jack.

## Collapsible Whip Preparation

File the plating off the small mounting stub at the base of the RadioShack 72-inch collapsible whip antenna. Once the bare brass is exposed, tin this with solder. Now insert the whip antenna base into the 1/8-NPT × 1-inch brass nipple such that the antenna base is just below the lip of the nibble. Temporarily hold these pieces together with some masking tape. Now heat the nipple with a soldering iron and solder the brass antenna base to the inside of the nipple. See Figures 5 and 6. Incidentally, I did find that some brass nipples were a little small on the inside to pass the collapsible whip. Nipples I purchased at True Value Hardware cleared the whip, and nipples purchased at Home Depot did not. If you can't find a nipple with a 0.275-inch ID, you can easily drill it out with a 9/32-inch drill bit.

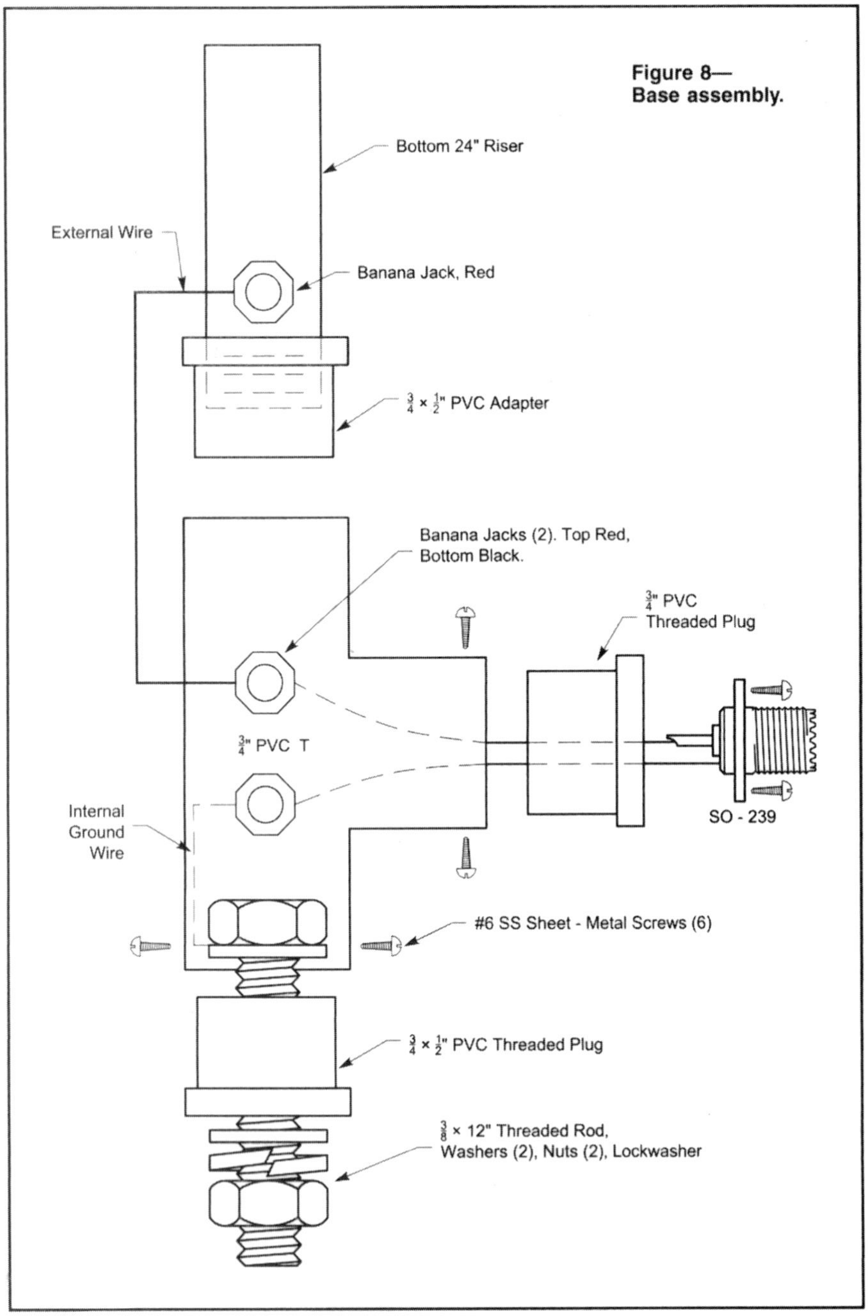

## Loading Coil Assembly

Cut off a 5-inch length of the Minductor 4027 coil. Unfold about half a turn from each end of the coil. On one end of the coil, solder a 6-inch piece of insulated wire terminated with an alligator clip. Using a screwdriver, indent every other turn of the coil. Finally, solder the coil leads to the brass screw heads on the 6-inch riser (adjust the brass screw lengths as necessary). The coil end, with the clip lead soldered to it, should be on the end of the riser that has the brass nipple showing. See Figure 7.

## Base Assembly

The base can be built easily just by referring to Figures 8 and 9. First, drill a 3/8-inch diameter hole into each of the ¾-inch PVC threaded plugs (one plug will be used for the ground support 3/8-inch threaded rod, the second plug will be used for the SO-239). Note that the plugs are threaded and the T is smooth. I used threaded plugs since they slip easily into the T. They will be held in place with #6 sheet metal screws. I also cut off about half of the threaded part of these plugs so as to leave more room inside the PVC T for the wiring. Place the SO-239 temporarily over the 3/8-inch hole just drilled in one of the ¾-inch threaded plugs and mark the location for two #6 stainless steel machine screws which will hold it in place. Also mark four points on the T as shown for the #6 stainless steel screws that will hold the two ¾-inch threaded plugs in place. Before finally connecting the SO-239 to the base, solder wires to the center conductor and to the ground for connecting to the internal connections as shown. You will probably need to file out a little more of the hole in this threaded plug so as to easily pass the ground wire. I used a short piece of copper braid from a piece of RG-58 cable from the SO-239 ground to the black ground banana jack, and soldered the end directly to the 3/8 × 16 brass nut. Drill 1/16-inch diameter holes for the six stainless steel sheet metal screws and complete the assembly of the base. I rounded the end of the 3/8 × 16-inch brass rod with a file to make pushing the base into the ground a little easier. I didn't sharpen the end, since this could cause a problem if you are trying to carry this through airline security!



Figure 8—Base assembly.

Figure 9—Base, prior to final assembly.

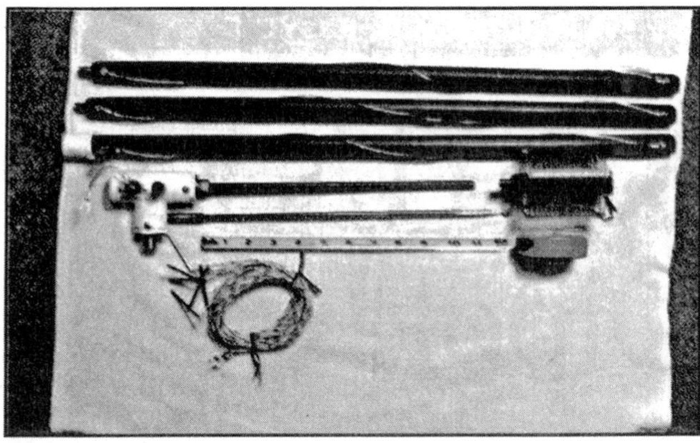

Figure 10—All of the antenna components.

Finally, attach ¼-inch spade lugs to each end of a 2½-inch long piece of bare #14 copper house wire. This will be the piece of wire that is used for attaching the base assembly to the bottom section of the antenna.

### Ground Radial Network

To minimize your ground losses, you really need some form of ground radials. These radials should be at least as long as the antenna is high. Therefore, I made six 12-foot radials using #22 insulated wire. Almost any gauge wire, insulated or not, can be used here. In my case, I attached all the wires together and to a ¼-inch spade lug on one end. This lug will attach to the lower black banana jack on the base assembly. You may wish to have six separate ground wires with their own spade lugs as unraveling the six folded-up radials takes longer than putting the rest of the antenna together. On each outer end of the radials, I soldered a 1-inch long piece of brass ⅛-inch rod. These ends are then pushed into the ground to help hold the radials in place.

The entire antenna, broken down into individual pieces, can be seen in Figure 10.

### Antenna Assembly

To assemble the antenna, first press the bottom antenna section (24-inch riser with banana jack) into the top of the base assembly and attach the 2½-inch length of interconnecting wire between these two red banana jacks. Then screw one of the remaining 24-inch risers into the top of the bottom antenna section (finger tight). Push this base/riser assembly firmly into the ground, keeping it as vertical as possible.

Now, assemble the remaining 24-inch riser with the loading coil assembly and the collapsible whip. Again, finger tight is all that is necessary for all brass fitting interconnections. Extend the whip, and screw this entire assembly into the open end of the 24-inch riser that is available on the assembly pushed into the ground.

Finally, extend the six radials, and attach the common end to the bottom black banana jack on the base assembly. See Figure 11.

Figure 11—Phil, AD5X, and the finished antenna.

Figure 12—Finding the permanent adjustment points on the coil for each band.

### Initial Antenna Setup

The idea here is to find permanent adjustment points on the coil for each band. So, starting with 40 meters, use an antenna analyzer to find the coil tap that gives the best SWR. See Figure 12. Mark this tap point. Move to 30 meters and repeat. Repeat again for 20 and 17 meters.

For 15, 12 and 10 meters, the entire coil will be shorted out and the top whip will be adjusted for resonance in these bands. I marked the top whip with a permanent black marker at the points necessary for these bands.

Now pull the loading coil/collapsible whip top assembly off, and solder short pieces of wire to the tap points determined for the 40 through 17 meter bands. From this point forward, you can just go back to these tap points, or adjust the top collapsible whip, and not have to worry about making SWR measurements. You'll find that in all cases, the SWR is under 1.5:1.

### Conclusions

I've described an inexpensive, yet efficient, portable vertical antenna for your operating excursions away from home. It assembles in minutes. While it is easily packed in a small travel bag or suitcase, you'll be amazed at the performance of this antenna. And, you'll have the pride associated with knowing that you built it yourself!

*Photos by the author.*

*Phil Salas, AD5X, has been a ham for 38 years and is an ARRL Life Member. He holds a BSEE from Virginia Tech and an MSEE from Southern Methodist University. He is currently Director of Hardware Engineering at Celion Networks in Richardson, Texas. At home he shares his station with his wife Debbie, N5UPT, and daughter Stephanie, AC5NF. You can contact Phil at 1517 Creekside Dr, Richardson, TX 75081-2913;* **ad5x@arrl.net**.

By Robert Victor, VA2ERY

# The Miracle Whip: A Multiband QRP Antenna

Want to hold the world in the palm of your hand? Tired of packing a suitcase-size antenna for your hand-held, dc-to-daylight transceiver? The Miracle Whip, a self-contained wide-range antenna made from inexpensive parts, can give you the flexibility you need to be truly free—no ground required!

One of my favorite radio fantasies started with Napoleon Solo—the man from U.N.C.L.E. He'd be in a tight spot, say, under fire from a crack team of THRUSH nannies in miniskirts, and he'd reach into his pocket and pull out the world's niftiest radio. It was about the size of a pack of cigarettes and had a two-inch whip antenna. He'd call up Control—who could be anywhere in the world at that particular moment—and try to muster some help. Control, of course, would dish out a number of droll comments about Solo's regrettable tendency to get into any number of tight spots, whereupon Napoleon would dial up partner Illya Kuryakin, on the other side of the room, and ask him to shoot back. The nannies, twittering like squirrels in a dog pound at having their pillbox hats punctured, would retreat in disarray. End of episode.

The mini rig was a prop, of course, and I realized even then that such a radio could never work. Short of satellite support (which would come soon enough) or a new understanding of the universe (which may or may not come), a two-inch whip on a hand-held HF transceiver might get a signal across a room, but not around the world.

Since then, often during evenings spent at a campground picnic table, I continued to think about what it would be like to have such a handy radio. I often visualized a book-size, multi-band rig powered by internal batteries; something that would be practical for cycling, hiking or working skip from any nearby picnic table. It was easy to imagine the rig rendered in such a portable package, but I could never get the two-inch whip to work—even in my mind. I guess I couldn't set aside *all* the laws of physics. Agent Solo (or his heirs and assigns) would be forever doomed to throwing wires into trees.

Although nobody seemed to consult me, the radio of my daydreams appeared on its own. When I saw the first magazine ads for Yaesu's FT-817 low power (QRP) transceiver earlier this year, I was delirious—it was *exactly* the rig I'd been fantasizing about. I dug out my credit card, told my wife I was ordering an Ab-Rocker and called my buddy Angelo at Radioworld in Toronto ("...but Honey, we can't send it back, we'll lose money on the restocking charge...").

Rig in hand, the man from U.N.C.L.E. was *still* in my thoughts. He wanted *his* radio, or at least something like it. He wanted an antenna that plugged into the back of his (my) new '817 so he could easily brandish it when in desperate need, without having to find a tree, when there were *nannies*. I tried to explain about antennas, but he merely gave me that pained, condescending look usually reserved for conversations with Control.

What might actually work here? A telescoping whip perhaps, around 50 inches long, with some kind of loading system so the antenna could cover all the HF bands. I'd have to stay away from "interchangeable" coils (Solo wouldn't want the hassle), and I'd have to produce some kind of workable results. Efficiency might be measured in the single digits on some bands, yet DX *had* to be a possibility.

What I came up with is definitely fit for an U.N.C.L.E. operative. It's a 48-inch telescoping whip with a homebrew loading and mounting device. Physically, it's portable and practical, and looks *secret agent cool* on the Yaesu. I finished construction just as a contest weekend was starting, so I got to try it out under ideal conditions.

Although my QRP signal didn't burn out anyone's receiver, I'm pretty satisfied with the results. I spent about fours hours on HF during this particular contest and had scads of contacts on 10, many on 15 and 20, and a couple on 17—almost all overseas! I also worked four stations on 40 (within about 400 miles) and managed one contact with a local operator on 80 meters. The rig was sitting on my desk—indoors—and the whip was plugged into the back of the radio, which was ungrounded. That's definitely a worst-case scenario! Because I figured it would take a miracle for a rig-mounted antenna to work DX, I christened my creation the "Miracle Whip"!

## In Theory

The heart of this design is in the loading system, which is made from readily available parts and costs about $30 for the whole works (less if you have the proverbial well-stocked junkbox). Here's the theory...

There are three ways (that I can think of) to load a length of wire on a particular frequency. The first is to make the wire a quarter of a wavelength long, which makes it resonant at the desired frequency. This works because the feed point impedance of a quarter-wave wire (assuming you have a counterpoise) is about 50 , which matches the coaxial output found on most rigs. Unfortunately, the shortest wavelength I'd be using was 10 meters, and a quarter of that is about eight feet, so this method wasn't an option.

The second way is to place a loading coil somewhere along the length of a wire that's shorter than a quarter wavelength at the desired frequency. You can place the coil at the base (base-loaded), somewhere in the middle (center-loaded) or at the top (end-loaded). Very simply, the loading coil makes up for the "missing" wire and forms a resonant circuit at the desired frequency.

How does it work? If you graphed the impedance of a quarter-wave antenna along its length, you'd see a continuous curve, with a low impedance at the feed end and high impedance at the far end. If you can imagine removing a section anywhere along the length of the antenna, you'd create a gap in that curve. The loading coil performs the impedance transformation required to bridge the impedance "gap" created by the missing section, allowing the use of a physically smaller (shorter) antenna.

A third method of achieving an impedance transformation is by using a transformer instead of a coil. A transformer is, after all, a device for matching different impedances! The hitch with this technique is that a transformer, unlike a loading coil, isn't a series device; it needs to be fed in parallel and usually "against" the antenna ground. Because of this factor, transformers must be used at the feed point.

Because Napoleon wouldn't like to swap loading coils to change bands, method three would have to be used. I figured an adjustable loading device would have to be placed at the base of the antenna, anyway, so a transformer seemed like a good possibility.

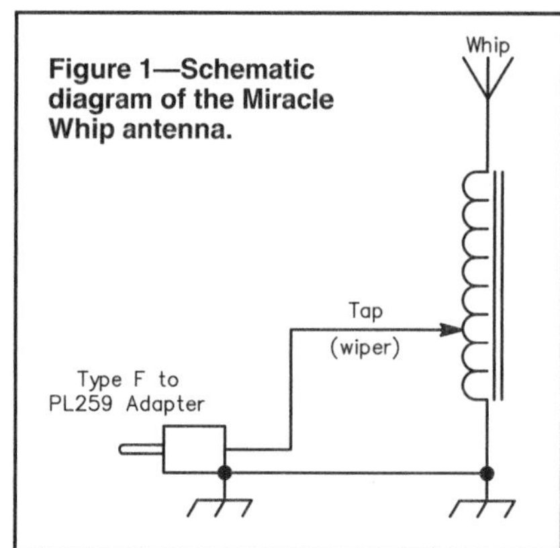


**Figure 1—Schematic diagram of the Miracle Whip antenna.**

**Close-up view of the inside of the Miracle Whip clearly showing the core and wiper.**

## The Autotransformer

Most of us are familiar with the broadband transmission-line transformers often used as baluns. They can be made somewhat adjustable with clever switching arrangements, but they're always limited to whole-numbers-squared ratios such as 1:1, 4:1, 9:1, 16:1 and so on. I worried that this limitation wouldn't allow for enough adjustment flexibility.

Thankfully, there's another kind of broadband RF transformer that *can* perform a match like this—the autotransformer—and it isn't limited to natural-square ratios. Although it's theoretically not as efficient as a transmission-line transformer, in practice it works quite well. The efficiency usually suffers as you apply more power (because of core losses), but at QRP power levels (5 W or less), those losses are minimal. With a little seat-of-the-pants engineering I came up with a way to make an autotransformer more-or-less continuously variable, which was exactly what I needed to use the same whip and matching unit over such a wide range of frequencies.

An autotransformer works like a conventional double-wound transformer as shown in Figure 1. The bottom part, where the input connects, represents the primary, and the entire coil, with the whip on the end, acts as the secondary. The impedance transformation is the square of the ratio between these two virtual sets of windings (turns). As the slider moves it taps the transformer, varying the ratio between the primary and secondary, providing (hopefully) the right match on each band.

This arrangement looks a bit like a series loading coil with a sliding tap, but if you look closer you'll see that we're applying the signal *across* the coil, which is connected to the signal source and to ground. The antenna winding—in effect the whole winding—is also across the output (the whip) and ground. Thus, we really do have a transformer as opposed to a loading coil, and the device does indeed transform our feed impedance into our whip impedance in a variable manner.

If you'd like to do a thought experiment, imagine exchanging the signal source and the ground, putting the ground on the tap and the source at the bottom. You'll see that the ratio is now different for any given tap position because the ground is now farther up the coil, which changes the number of windings on the antenna side. You'll also see that you've reversed the phase of the output. If you build and test this, you'll confirm this result.

## Construction

I'm no machinist, so it was challenging for me to figure out how to homebrew the mechanics of the Miracle Whip. When I have no idea how to create what I need, a wander through the local surplus shop will occasionally provide inspiration.

I did just that, and happened to find a wire-wound rheostat that looked like it was designed for just this project. It had the perfect wiper-and-brush mechanism that I'd need to make the sliding tap, and the resistance winding and the coil form it was wound on looked a lot like a toroidal transformer, which gave me some confidence that the unit could be adapted for my needs. It worked well, so here's how to build your own transformer out of a similar rheostat.

I've located some common commercial rheostats made by Ohmite that you can order from any of several suppliers. Go to the Ohmite web site at **www .ohmite.com**, click on "distributors" and choose one near you (or order from the Allied site in the parts list). These rheostats are supplied in many resistance values, but because you won't be using the resistance winding you can take anything that's in stock that's the correct physical type. These are identified as Ohmite part number RES*xxx*, with the "*xxx*" being the resistance. Typical values are shown in the parts list.

I'm going to go into quite a bit of detail on the construction of this device, but don't be intimidated—the whole process is straightforward and shouldn't take more than a couple of hours.

Start building by stripping the rheostat. You'll use the central shaft, which has a spring-loaded wiper and brush, its associated hardware and the collar/tube in which the shaft rotates. You can toss the resistance winding into your junk box. To get these parts free you'll need to unscrew the collar-retaining nut and remove the C-clip that holds the shaft in the collar. Don't lose the C-clip and be careful not to stress the wiper spring and its contact. The brush is held in its seat on the wiper by pressure alone, so when you take it apart, expect the brush to dangle on its pigtail.

## Winding the Transformer

The transformer (Figure 2) is created by winding about 60 turns of #26 enameled wire onto the ferrite core specified in the parts list. I say "about" 60 turns because the number of turns isn't critical. A loading coil would need exactly the right number of turns on exactly the right core for consistent performance, but because our device is a broadband transformer, we're only concerned with the appropriate ratios between the primary and the secondary. Because the windings ratio of the finished unit will be adjustable anyway (that's why we're building it, right?), the number of windings isn't overly critical.

That said, you should shoot for about 60 turns; one or two less or more shouldn't be a problem. What you *do* want are uniform windings that are tight on the core, regularly spaced, with a bit of room between the windings (so the brush will contact only one at a time) and a gap of 30 degrees or so where there are no windings at all.

Why the gap? The rheostat, as originally manufactured, has stops to prevent rotation beyond the ends of the windings, but we lose those stops when we discard the original mounting. The gap will give you a good "feel" for when you've reached the beginning or end of your windings as you tune, so you'll know where you are. (If you think of a better solution, let me know.)

Spread some non-corrosive glue (Elmer's wood glue works fine) on the bottom and rim of the core to hold the windings in place and let the assembly dry completely before proceeding. Use a piece of fine sandpaper or emery cloth to carefully remove the enamel from the wire in the area where the wiper will make contact. You can eyeball this area by temporarily placing the wiper on the core with the shaft centered through the hole in the core.

## Mounting

Here's the only tricky part of the project—mounting the core, the wiper and the shaft so the wiper contacts the coil windings with a suitable pressure. If the wiper is too high above the windings, you won't get good contact; if it's too low, adjustment will be difficult and you might tear the brush and perhaps even the windings. That said, it's not *that* difficult to get this right. Look at Figure 3 to understand the mechanics.

Cut a square of perfboard about $1^1/_2$ inches to a side and drill a hole dead center to accept the shaft collar. Center your newly wound core over the hole. Slide the wiper and shaft into the collar and install the C-clip. Insert the wiper/shaft/collar assembly through the core and the hole in the perfboard with the wiper positioned to contact the windings. Pull on the shaft and collar from the opposite side of the perfboard to see how things fit. If the shaft collar flange bottoms out on the perfboard *and* the wiper is contacting the windings with a reasonable-but-not-excessive force (there's still some spring travel in the wiper), you're home free.

If the wiper spring is bottoming out before the shaft collar flange is firmly seated on the perfboard, you'll need to insert one or more washers between the flange and the perfboard until the fit is right. This happened to me, and I wound up cutting a washer from a piece of transparent Mylar to get a good fit.

On the other hand, if the shaft collar flange bottoms out on the perfboard but the wiper contacts the windings only lightly (or not at all), you'll need to elevate the core above the perfboard by shimming underneath the core. You can do this by cutting a core-shaped ring of glueable, non-metallic material that's the right thickness, and gluing it under the core to raise it enough to get good contact between the wiper and the windings.

Fortunately, the wiper spring has a good deal of travel, so this adjustment isn't too difficult. Don't rush it, however, and spend enough time to set this up properly.

Once the adjustment's right, glue the core permanently to the perfboard, centering it over the hole and set it aside to dry. You can then insert and fasten the mounting collar with its nut. Finally, remove the C-clip from the shaft and extract the shaft and wiper for the next step.

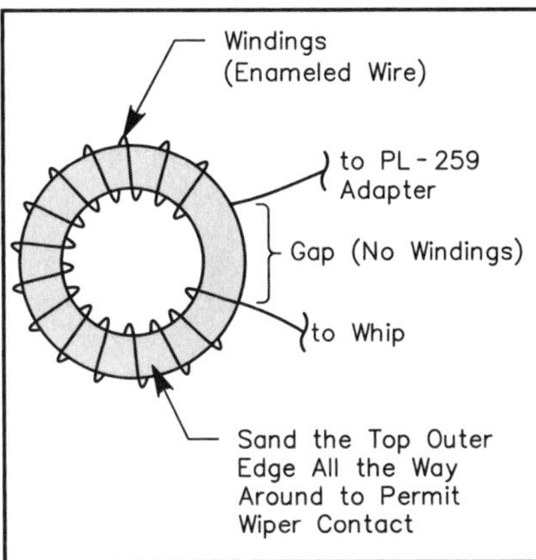


**Figure 2—Winding the ferrite core with approximately 60 turns of #26 enamel wire. Note that you must sand the windings along the top outer edge to remove the enamel coating so that the brush can make contact.**

### Parts List

- Wire-wound rheostat—Ohmite # res100, res250, res500, res1000 or similar (available from Allied Electronics, **www.alliedelec.com**, about $20 each).
- Core—Palomar F82-61 or similar (available from Palomar Engineers at **www.palomar-engineers.com**; about $1.60 each).
- Whip, wire, PL-259, etc
- Enclosure—Hammond #1551HBK or similar.
- F-female to PL-259 adapter—RadioShack 278-258.

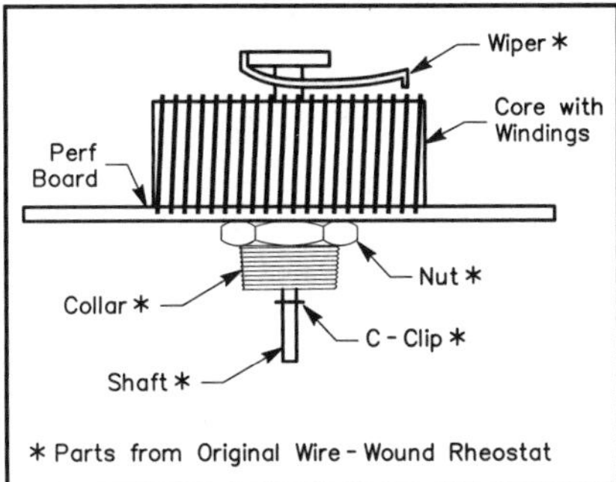


**Figure 3—Side view of the modified rheostat assembly. The wiper and brush make contact with the core windings, jumping from one winding to another as you turn the wiper shaft.**

## The Brush

The original brush is quite wide for our purposes, so we need to file it down so it forms a flattened point that will contact only one winding at a time. You're going to file the sides and top to shape the contact area like a wedge with a flattened point. Check out Figure 4 to see what I mean.

Use a fine-tooth file and go slowly. The brush material is quite soft and you don't want to go too far. After shaping, use the file to cut a shallow groove across the middle of the point. This helps the point seat solidly when it settles over a winding. Make sure to round the edges as shown so the brush doesn't hang up when stepping over the windings.

After this you're ready to insert the shaft and wiper into the collar and replace the C-clip on the shaft to hold it in place.

## Assembly

All that remains is to install your completed transformer assembly, a PL-259 coaxial connector and whip in a suitable enclosure. The transformer unit and the coaxial connector should be mounted so they don't interfere with each other, and the whip mounts on the top of the box. Eyeball the positions before you drill any holes. That done, drill all three required holes in the appropriate locations.

Panel-mount PL-259s are few and far between, but I managed to find something suitable. It's an "F-female to PL-259" adapter sold by RadioShack, part number 278-258. There's no way to solder to the inside (the F-type end of this adapter), but it's designed to make good contact with a piece of solid wire inserted straight into the end (like a cable-TV connector), so cut a short length of solid copper hookup wire, remove the insulation and stick it in the hole. You'll solder a lead from this to the transformer wiper lead. Your ground connection can be provided by using an appropriately sized lug washer (if you can find one) or by slipping another stripped lead under the connector nut as you tighten it down.

Mount the transformer in the box by inserting the shaft collar through the mounting hole and install the retaining nut.

My whip is 48 inches long and came from a surplus store. It looks like it might once have been part of a "rabbit ear" assembly. I chose it because it's beefy and

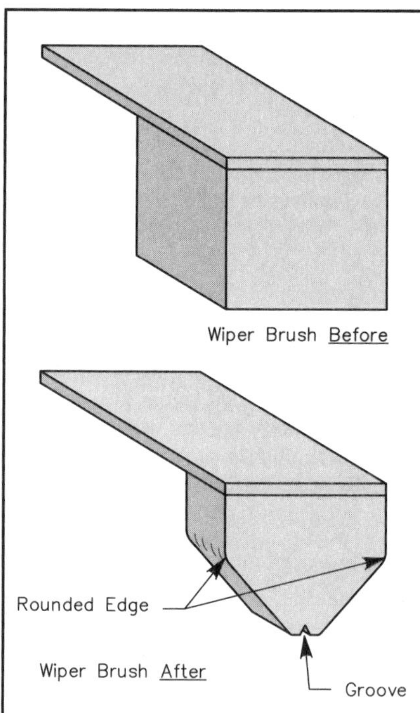


**Figure 4—The normally flat-edged wiper brush must be gently filed to a rounded point (with a narrow groove) and rounded corners.**

because it had a swivel mount that would allow swinging the antenna to a horizontal or vertical orientation. Mount yours to the top edge of the box, making a connection in whatever fashion required; a stripped lead or lug under the mounting screw should do just fine.

Wire things up as per the diagram and remember to use a thin, flexible lead to make the connection between the wiper the Type-F end of your PL-259 adapter. Make sure there's plenty of slack. You'll want this to move freely, without strain. Screw on the cover and plug it in!

## Operating

Select a band and tune the antenna by rotating the wiper while listening to band noise or a signal. The antenna peaks nicely on receive, so if you don't hear something right off the bat, something needs to be checked. You may find that the whip works better horizontally or vertically. Listen and experiment to determine how the antenna performs with the station you're working.

Once peaked for maximum receive signal, transmit at low power while watching the FT-817's SWR meter. If you have significant reflected power, rotate the slider a little to one side or the other and try again. You can feel each "detent" as you step from winding to winding. You might not get a perfect match on the lower bands because the impedance transformation ratios jump rather quickly at the bottom end of the transformer, but you should get something that's workable. I get 1:1 on 20, 15 and 10, and about 2:1 on 40 and 80 meters. Remember that your transmission line is about two inches long, so SWR-induced line losses aren't a consideration—you're mainly looking for reasonable loading.

A few words to the wise: *always tune at the lowest power setting and never attempt to transmit at higher power unless you see a decent match.* And, as mentioned before, the antenna peaks nicely on receive, so if you don't hear a peak, investigate and fix things before you transmit!

Once peaked, you're ready to switch to higher power and talk to someone. Remember that you're working QRP *with a compromise antenna.* A little patience will go a long way and, like a glider pilot or a fisherman, waiting for the right conditions is half the battle.

## Performance

I'm not sure this setup would have saved Napoleon Solo's bacon every time, but considering the challenges of operating a QRP rig with an attached whip, I'm very pleased with the results. I've made the contacts I described with the whole kit and caboodle sitting on my desktop, without any sort of ground or counterpoise. In fact, adding a ground might make impedance matching considerably more difficult.

Obviously, the antenna performs better at higher frequencies. On 10 meters it's about an eighth of a wavelength—which isn't bad. As you go down in frequency the antenna is electrically shorter and less efficient. But it loads and radiates all the way down to 80 meters, which is the design goal, and it *will* make contacts there, given the right conditions.

## Six, Two and More and More

Although I didn't design it to do so, the antenna works great on 6, 2 and even 440. The trick is to set the wiper to the very last turn—in effect providing a direct connection to the whip—with the transformer simply acting as a choke to ground. You can then slide the whip in or out to approximate a quarter wavelength for whatever band you're on. In this case the antenna is full size, so there's no compromise at all!

**Exterior view of the Miracle Whip housing.**

The autotransformer principle should also be applicable to a general-purpose, random-wire tuner. I think I'll play around with this. If you're an intrepid experimenter, I invite you to do the same and let me know what you find.

This antenna should work with just about any QRP rig, homebrew or store-bought. The only proviso is that, although the TX outputs of almost all rigs are designed to work into 50 Ω, the receiver inputs may prefer other impedances. Receiver input impedance is far less critical for most applications, however, so this may not be much of a handicap.

I'm completely satisfied with my first Miracle Whip—so much so that I plan to offer a commercial version to the amateur community (on the Web see **www.miracleantenna.com**).

With the Miracle Whip I've realized a radio dream I've had for many years: working the world with a self-contained, hand-held station. I haven't yet tested it on a picnic table, but my desk is a pretty fair substitute. I'm expecting Napoleon to knock off the condescension once we get out to the campground.

The Miracle Whip trades efficiency for size and portability, so don't expect, well, miracles. But if you want a system that can work DX from a picnic table, an ocean view or a mountaintop, this one does the trick. Now, Mr. Solo, about those nannies…

*You can contact the author at 1220 Bernard St, No. 21, Outremont, QC, H2V 1V2, Canada;* **Lebloke@attcanada.ca**.

# Build an HF *Walking Stick* Antenna

**By Robert Capon, WA3ULH**
**322 Burlage Circle**
**Chapel Hill, NC 27514**
**Photos by the author**

This antenna project was inspired by the planning that went into an expedition by a group of amateurs to activate Harker's Island off the coast of North Carolina. We had just finished operating the 20-meter CW station with the Orange Country Radio Amateurs during Field Day. We had so much fun, we didn't want to wait another year to combine ham radio with a camping adventure.

As with any expedition, one of the central planning questions is, "What do we use for an antenna?" During Field Day, we worked QRP, so we opted for the gain of a 3-element Yagi mounted on a 21-foot mast, held with guy ropes and massive anchors. The mast weighs about 50 pounds, so setting up the Yagi was hard work. Fortunately we had six sturdy hams on hand to do the job safely. But we had only four amateurs for the IOTA expedition, and none of us were excited about setting up the Yagi.

None of us are big fans of dipoles. Dipoles have a rather high angle of radiation, and have nulls that are very inconvenient. And dipoles can be difficult to put up on a flat, open beach.

This left us with our commercial multiband verticals. I like my Cushcraft R-7 and my friend Paul, AA4XX, loves his GAP. With 100 W, we felt that we could work the whole world with our verticals. In fact, I started my DX career with only 100 W and a ground-mounted Butternut HF6V vertical. I worked over 200 countries before I graduated to a Cushcraft A-3S Yagi.

But our plans for the mini-expedition called for spending most of our time on only two bands: 20 and 40 meters. So why not put together single-band vertical antennas held up with lightweight guy ropes? This would enable us to leave our commercial verticals at home, thus eliminating the need to dismantle our home stations.

With a little effort, I designed a monoband vertical antenna that can be fit into a walking stick that weighs about 3 pounds. The antenna can be fabricated in less than two hours, assembled in the field in less than 15 minutes (either free-standing or suspended from a tree), and built for about $30. Not bad, considering that regular walking sticks found in trail shops run about $70!

In addition to use as an expedition antenna, the walking stick vertical is an excellent monoband antenna for the new ham. It provides performance on par with commercial vertical antennas, except that it only works on a single band and it's very inexpensive. The antenna is ideal for both QRP and full-power operation.

Before we proceed, please keep in mind the cardinal safety rule when working with vertical antennas. Do not allow the antenna to come into contact with a power line or power source of any kind—you could be killed. That's why commercial verticals carry a precautionary warning label. Please use caution whenever you are in the vicinity of electrical outlets, fixtures and overhead power lines.

### Quarter-Wavelength Verticals

The walking stick is a design variation of the classic quarter-wave vertical antenna. The design is optimized when a vertical quarter-wavelength radiator is combined with a tuned quarter-wavelength radial system.

The vertical element can be as simple as a quarter wavelength of wire or ladder line, suspended from a tree using nylon rope and an insulator. Three to four wires, also trimmed to a quarter wavelength, work very well for the radial system.

To determine the length in feet of a quarter wavelength, use the following formula:

Length (in feet) = 234 / f(MHz)

This works out to the dimensions shown in Table 1. You'll notice that the lengths for the CW and phone portions of each band are only a few inches apart, so the same antenna can be used for different parts of the band without adjustment. Further, the same antenna can be used for any of the higher bands by compressing its size during assembly in the field.

**Table 1**

| *Band* | *CW (ft)* | *Phone (ft)* |
|---|---|---|
| 10 meters | 8.3 | 8.2 |
| 12 meters | 9.4 | 9.4 |
| 15 meters | 11.1 | 11.0 |
| 17 meters | 12.9 | 12.9 |
| 20 meters | 16.7 | 16.5 |
| 30 meters | 23.1 | — |
| 40 meters | 33.3 | 32.4 |

A quarter-wavelength vertical exhibits excellent characteristics for working DX because it has a very low angle of radiation. The antenna has a low impedance at the feedpoint, it is omnidirectional, and it exhibits very broadband coverage for the band of interest.

Unfortunately, even multiples of a quarter-wavelength vertical do not work well because they have a high impedance at the feedpoint. Therefore, a quarter-wave vertical on the CW portion of 20 meters that's a half wave on 10 meters will exhibit unacceptably high impedance on 10 meters (the SWR will be high).

Odd multiples do resonate, but their angles of radiation increase to the point where they become much less effective. For example, a quarter-wave vertical on 40 meters is approximately $^3/_4$ wave on 15 meters and will resonate on 15. But at $^3/_4$ wave, the antenna has a high angle of radiation and is much less effective for working DX on 15 meters. So the magic number for the monoband walking stick vertical is a quarter wavelength.

### Parts List

All of the parts that you will need are readily available, and the total cost of the antenna is about $30 (see Table 2).

I was surprised at how difficult it was to locate telescoping (nested) aluminum tubing at my local hardware stores. The $^5/_8$-inch tube was impossible to find. Fortunately, my DX buddy Ernie, AD4VA, came to the rescue by suggesting that I contact a mail-order supplier. Sure enough, I was able to order the four lengths from a mail-order supplier for $16.50, plus shipping.

When the aluminum arrived, I found that two of the sections did not fit together at all, and I had to swap the $^3/_4$-inch tube with a section from the hardware store to build the antenna. My friend Paul, AA4XX, who built a 40-meter walking stick vertical, encountered the same problem. In fact, two of his sections became stuck together and could not be separated.

Be careful. The reason that the pieces can fit improperly or become jammed is that aluminum suppliers ordinarily cut aluminum to length without further preparation of the ends. Before you attempt to nest the tubing, you must bevel and deburr the ends with a rounded metal file, and carefully clean away the aluminum filings. To clean away the filings, place a wad of soft cotton cloth sprayed with WD-40 in each tube, and push the oiled cloth through the length of the tube (like a ramrod) with the smallest diameter tube that you have.

Remember, the aluminum sections have extremely tight tolerances. If a single particle of aluminum remains inside an aluminum tube, the tubes will likely jam.

### Building the Aluminum Sections

I spend most of my time on 20 meters, so I built my walking stick vertical for that band. This choice also enables me to compress the antenna during assembly in the field to work on 10 through 20 meters as the need arises.

Begin constructing the antenna by cutting the aluminum tubes to the proper lengths with a hacksaw. The aluminum tubing cuts easily, but please be sure to wear safety glasses. Cut the three smaller diameter tubes to a length of 4 feet, 5 inches. Cut the 1-inch tube to a length of 4 feet, 8 inches.

Next, prepare the antenna for assembly by cutting 2-inch slots in one end of the 1-inch, 7/8-inch, and 3/4-inch tubes. (The 5/8-inch tube does not need a slot.) To slot the tubes, gently place them in a wooden vise, and cut the slot with a hacksaw. Be sure not to make the vise too tight, because this will compress the tubing, and make it harder to nest the tubes later. After cutting the aluminum, gently file off the burrs with a fine-grade, curved metal file and clean away the filings.

When you assemble the antenna in the field, place the hose-clamps over the slotted ends of each of the aluminum tubes and attach the unslotted tube of the next smaller size. (Slide the 7/8-inch tube into the 1-inch tube, and so on.) Gently tighten the hose clamps for a snug fit.

Measure the length of the vertical for the band of interest, and label the three smaller tubes with a permanent laundry pen or a scribe mark to facilitate setting up the an-

**Table 2**

Note: The lengths shown are for 10 through 20 meters

| | |
|---|---|
| Aluminum Tubing: (use 0.058" wall thickness for proper telescoping) | 1-inch×72 inches<br>7/8-inch×72 inches<br>3/4-inch×72 inches<br>5/8-inch×72 inches |
| Guy ring | Scrap of wood or closet bar hanger |
| Guy rope | Nylon, 1/8 inch or braided mason line, 60 feet |
| Tent stakes (3) | Lightweight stakes |
| Radial wire | 50 feet insulated #22 or heavier |
| Hose clamps (3) | 1 1/4-inch hose clamps |
| Rubber walking tip | Cane tip (purchased used from medical supply store) |
| Rubber antenna tip | Rubber chair tip, from hardware store |
| Hand grip | Black cloth tape or tennis racket grip |
| Decorative knob | 2-inch diameter wooden sphere |
| Ground radial post | 1/2-inch-diameter PVC or fiberglass tube |
| Fishing Weights (4) | 1/2-oz brass fishing weights |

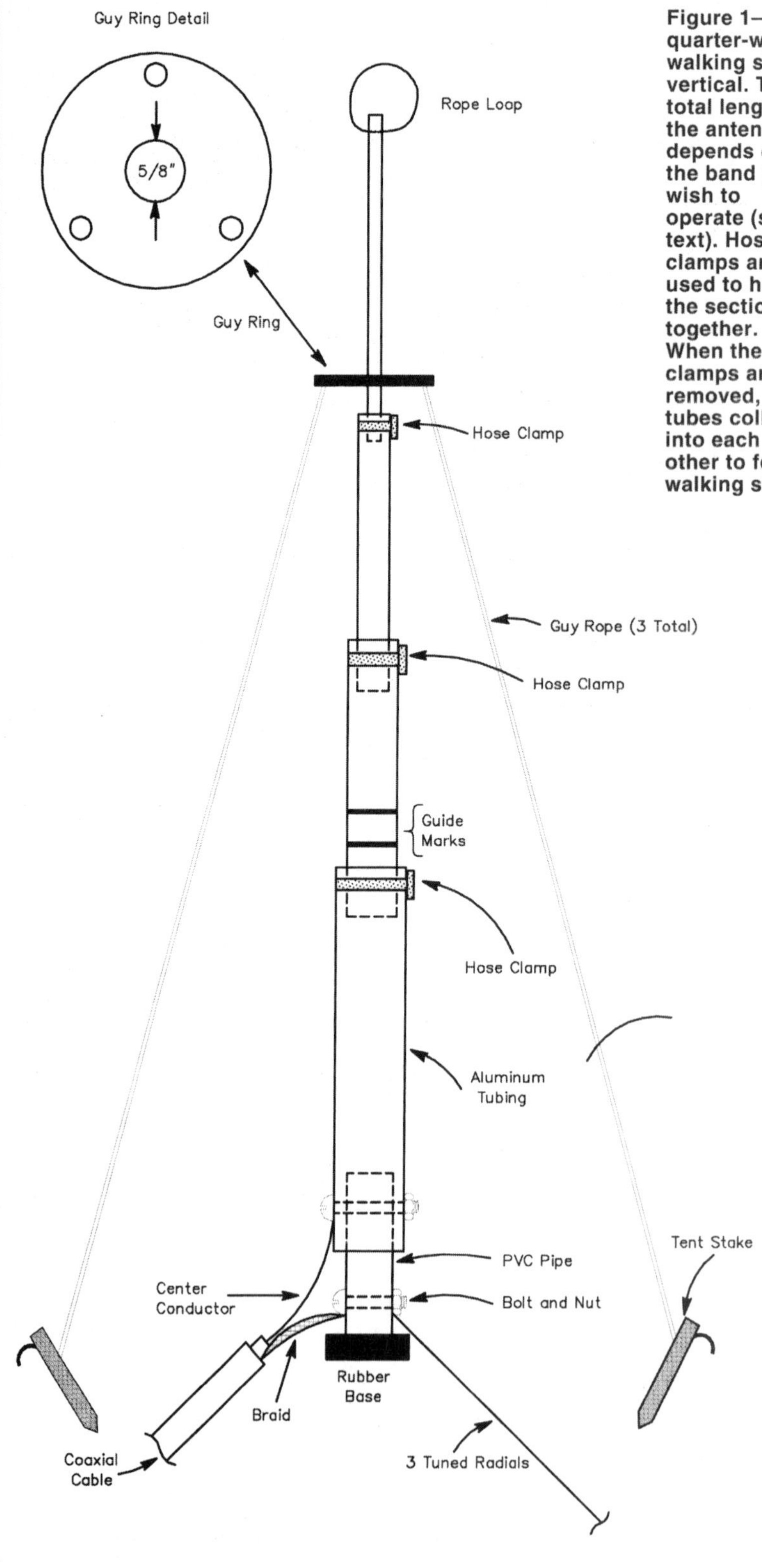

Guy Ring Detail

5/8"

Rope Loop

Guy Ring

Hose Clamp

Guy Rope (3 Total)

Hose Clamp

Guide Marks

Hose Clamp

Aluminum Tubing

Tent Stake

PVC Pipe

Center Conductor

Bolt and Nut

Rubber Base

Braid

Coaxial Cable

3 Tuned Radials

**Figure 1—The quarter-wave walking stick vertical. The total length of the antenna depends on the band you wish to operate (see text). Hose clamps are used to hold the sections together. When the clamps are removed, the tubes collapse into each other to form a walking stick!**

tenna in the field. I marked the tubing for the middle of the 20-meter band. I also marked the 7/8-inch tube for several of my other favorite bands. (The 7/8-inch section is ideal for length adjustments because you can reach it from the ground.)

Finally, drill a 1/8-inch hole approximately 1/2 inch from the top of the 3/4-inch aluminum tubing. Thread a small piece of nylon rope through the hole and tie the rope into a small loop. This loop will come in handy in situations where it is more convenient to suspend the vertical from a tree limb, rather than erecting it as a free-standing antenna with guy ropes.

Install the rubber cane tip on the end of the 1-inch aluminum tube. I went to a medical supply store where the clerk advised me that new cane tips cost $9.30 a pair. I noticed that there was a box of assorted used cane tips for 30 cents each, and I opted for one of these. This tip is used when the antenna is in the "walking stick mode." You can also buy chair tips, which are not as sturdy but are much easier to find and cheaper to buy, in most hardware stores.

### Building the Guy Ring

To build the guy rope ring find a scrap disk of wood or heavy plastic, approximately 2 to 3 inches in diameter. Almost anything will work for this application. I used an item commonly found in hardware stores: a heavy plastic disk designed to hold a closet pole. Drill a 5/8-inch hole in the center of the disk. Then drill three 1/8-inch holes, equally spaced 120° apart on the perimeter of the disk (see Figure 1).

Tie three 17-foot lengths of nylon rope onto the guy rope ring, using a non-slip knot. I used braided mason "snap line" for my guy ropes to support this very lightweight antenna.

When you assemble the antenna in the field, slide the guy ring down the 5/8-inch aluminum rod. The ring is held in place by the 3/4-inch tube, so the guy ropes will go approximately 3/4 of the way up the antenna.

### Building the Ground Radial System

Cut three ground radials for the band of preference to the same length as the antenna. Figure 2 shows the detail for connecting the ground radials to the antenna. The center conductor of your coax feed line goes to the vertical aluminum tubing, while the coax braid goes to the radials. Attach a small fishing weight to the end of each radial. This helps weigh down the radials in the field. These fishing weights are also commonly available in brass, which is much safer to handle than lead weights.

To insulate the ground radials from the center conductor, I used a 6-inch length of 1/2-inch inner diameter PVC plastic mounted with a 1/8-inch bolt to the bottom of the 1-inch aluminum tube section (Figure 2). I finished off the PVC section by installing a rubber chair tip on the end of the PVC insulator tube.

This PVC ground section is installed when the antenna is operational. To avoid damaging the PVC tube, remove the tube when you are back in "walking stick mode," and replace it with the rubber cane tip (directly mounted on the aluminum tube).

**Figure 2—Base detail of the walking stick vertical. The center conductor of the coax is bolted directly to the aluminum tube. The shield is bolted to the PVC pipe along with three tuned radial wires. The rubber tip at the bottom of the PVC pipe doubles as the walking stick base.**

### Finishing Touches

I used a wooden knob for the top of the antenna to give it the "Professional Walking Stick Look." I used a 2-inch-diameter wooden ball (an inexpensive cabinet pull from the local hardware store), and drilled a 1-inch hole in the bottom to form the knob. Finally, I stained and sealed the knob with tung-oil finish.

To put the final touches on the antenna, I fashioned a walking stick grip at the top. Black cloth electrical tape would be sufficient for the purpose, but I used a tennis racket replacement grip, which winds onto the tube like tape. (These replacement tennis racket grips can be found at almost any sporting goods shop that sells tennis accessories.) I also purchased a small nylon pouch to provide a convenient place to store the miscellaneous parts, guy rope, and ground radial wires of the antenna.

If you want a classy extra touch, consider putting a compass into the wooden knob. (My wife and kids love this feature.) I found a very nice compass built into a $2.99 survival kit at a department store. I pried the compass out of the plastic kit, drilled a shallow hole into the top of the wooden knob and cemented the compass in place.

*An antenna that gets you on the air and keeps you on the path.*

### Operational Test

In preparation for our expedition to Harker's Island, I assembled a team to go out to a local baseball field to set up the antenna and test it out: Ernie, AD4VA; my son Howard, KE4RUZ (age 7); and myself. We opted for the guy rope method of installation because we were anxious to try out the portable guy system.

My heart skipped a beat when we hooked the antenna up to the SWR analyzer and saw high readings on all frequencies! I got out my digital multimeter and began testing the coaxial feed line when Ernie and I simultaneously noticed that the center conductor of the coax had broken off. We fixed the connector and were in business.

Note that three on-ground radials don't make for a terribly efficient 1/4-wave vertical antenna, but our saltwater location helped even the score. The vertical will work even better if you can get the antenna base and radials 8 or 10 feet in the air.

I brought along my Index Labs QRP Plus 5-W rig and we put the antenna through its paces. Signals were loud and seemed to be coming in from every direction. In about 15 minutes of operation we logged two contacts: VE3DMC, in Ottawa, and W5RRR, the station at the Johnson Space Center in Houston. We made contact with both stations on the first call and received reports of 579 from each. I was running just 3 W at the time.

Paul's walking stick vertical for 40 meters delivered similar performance. His antenna had seven 5-foot sections, with the largest section made of 1 1/4-inch aluminum. Paul's 40-meter antenna was so tall, he used two sets of 1/8-inch guy rope to erect it. Paul tested his antenna on 40 meters and obtained an SWR of 1.3:1 or less over a bandwidth of 300 kHz. Paul tested the antenna for approximately 30 minutes running 100 W and worked five European stations including: EC5ABU, OK1DOZ, UA2FCC, and F9OQ.

### Next Steps

You may want to experiment with an assortment of mounting possibilities to set up the walking stick vertical for semi-permanent use at your home. After you have mastered the monoband walking stick vertical, you may also want to try fabricating a multiband vertical. *The ARRL Antenna Book* (Chapter 15) has a design for an excellent two-band vertical for 10 and 15 meters that is also inexpensive to build.

Thanks to my expedition buddies (Paul, AA4XX; Joe, KD4LLV; Gerry, KD4YJV, and Howard) for helping to inspire this antenna design. Also, special thanks to my field testers, Ernie, AD4VA, and Howard, KE4RUZ.

By Robert Johns, W3JIP

# A Ground-Coupled Portable Antenna

As the saying goes, "Imitation is the sincerest form of flattery." Here's a home-brewed antenna that proves the point.

This homebrew portable antenna for 40 through 6 meters is patterned after the ground-coupled design pioneered by Alpha Delta Commun-ications, Inc.[1] Instead of using radials, this antenna employs a simple and very small grounding system that needs no tuning.

The antenna described here is a quarter-wave vertical sitting on a tripod base. The vertical mast and the tripod are each made of 2-foot-long telescoping sections of 3/4- and 5/8-inch-diameter aluminum tubing.[2] The mast itself resonates on 10 mete lightweight aluminum tubing sections added to the top of the mast to tune the antenna to 12, 15 and 17 meters.[3] These added tubing lengths can be installed vertically or horizontally. The antenna is fed at the top of the tripod, making the base a part of the radiating system. A bungee cord stretched from the top of the tripod to a stake in the ground keeps the structure stable.

Beneath the foot of each tripod leg is a grounding strip 2 1/2 inches wide and about 3 1/2 feet long, made of aluminum tape.[4] These strips are simply laid on the ground and form one plate of a capacitor coupling RF from the antenna to the ground. That's the whole grounding system! When I read about this in *QST* (see Note 1), I was skeptical, but intrigued. The arrangement is similar to that of a mobile antenna system in which the car body acts as one plate of a capacitor coupling RF to the road and ground. This grounding system works: The antenna radiates well and the SWR is reasonably low on all bands. (The tripod and grounding strips can also be used with any vertical element or mobile whip you have.) A loading coil added between the aluminum tubing mast and the flattop permits operation on 20, 30 and 40 meters. With the coil positioned this far up the antenna, the entire 10 feet of tripod and mast are unloaded radiators on all HF bands.

## Building the Tripod

The top of the tripod, Figures 1 and 2, makes it easy to set up. The three 5/8-inch-diameter × 0.058-inch-wall aluminum tubes extending from the 1 1/2-inch PVC cap are permanently attached to it. To assemble the tripod, the legs slide over these tubes. A 3-inch-long, 3/8-inch carriage bolt passes through a hole in the top of the PVC cap to support the vertical element. This bolt also grips the 4-inch-long aluminum tubes inside the cap to form the three sloping legs of

[1]Notes appear on page 00.

and its caption for details on how to make this top cap.

A 50-Ω coaxial feed line attaches to the antenna via an SO-239 chassis connector mounted on an aluminum angle bracket at the top of the PVC cap (see Figure 1). Make a 5/8-inch-diameter hole in the bracket to accept the coax connector body; you'll also need to drill four small holes for the connector's mounting hardware. The 3/8-inch bolt through the top of the cap keeps the aluminum bracket in place.

**Figure 1—The top of the tripod with the bottom section of the mast connected to it. A bolt holds the three leg supports in the PVC cap slots. This bolt also passes through the 1 1/2 × 1 1/2-inch aluminum-angle piece that supports an SO-239 chassis connector for feed-line connection. A 1/4-inch hole in the cap top accepts the bungee-cord hook.**

To assemble the tripod top, invert the cap so that you are looking down at the open end. Insert the carriage bolt through the 3/8-inch hole in the cap, through the mounting hole in the aluminum angle and add a lock washer and nut to the bolt. Initially, thread the nut about an inch onto the bolt so that the bolt is still loose and its head is out of the cap. Insert the three aluminum tubes into the slots in the wall of the cap and down against the carriage bolt where it passes through the hole in the cap. Tighten the nut so that the carriage-bolt head squeezes the tubes outward and into the slots. Once the nut is hand tight, wriggle each tube to seat it snugly with its tip into the countersunk hole with the bolt. Tighten the nut until the round wall of the cap is slightly deformed into a triangular shape.

Each tripod leg consists of a 0.058-inch-wall, 3/4-inch-diameter tube and a 0.058-inch-wall, 5/8-inch-diameter tube that fits inside the 3/4-inch tube. Each tube is two feet long; three can be made from 6-foot tubing lengths. Dimple each 3/4-inch tube about one inch from each end. The dimple acts as a stop and prevents the smaller tube from penetrating any farther. Form the dimples using a couple of firm hammer taps on a center punch placed against the tube.

[1]Notes appear on page 42.

When joining the tubes, push a bit when inserting the smaller tube so that the side of the dimple holds the smaller tube in place.

## Ground Strips

Although mating two strips of aluminum tape with their sticky sides together might seem like a routine job, it's probably the most difficult part of building this antenna! The adhesive is quite sticky and unforgiving, and handling the long strips can be messy. Get an assistant to help you with this task. You'll need three strips.

See Figure 3. Cut a 7-foot length of tape from the roll and lay it down, sticky side up, on the floor or a large table. Have your helper press a piece of heavy (#12) solid wire or a thin dowel across the width of the strip at the 3½-foot midpoint and hold it in place. Pick up one end of the strip and carry it over the midpoint, keeping it tight so that it doesn't sag and touch the lower half. Keep both ends of the strip aligned while your helper at the midpoint presses the top piece of tape against the lower, working their way toward you. Trim (or remove) the excess wire or rod and the ground strip is done. Don't worry if the strips aren't aligned perfectly.

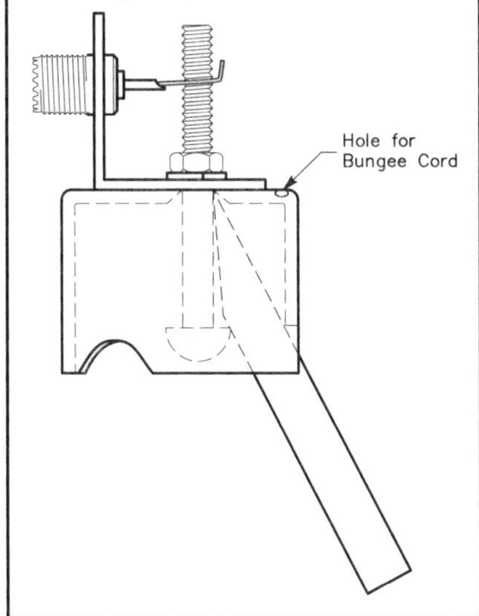


**Figure 2—The tripod top cap. The three ⅝-inch-diameter aluminum tubes are 4 inches long, cut at a 30° angle at the end within the cap. From inside the cap, countersink a ⅜-inch hole in the cap top. This forms a trap that holds the ends of the aluminum tubes. Although only one of these leg supports is shown, all three are held between the bolt head, the three slots (either carved or filed in the wall of the cap) and the countersunk hole in the cap top. The slots in the cap are about ⅜-inch deep and wide enough to receive the aluminum tubes. An easy way to lay out the slots is to use the fluted handle from an outdoor water faucet as a template. The handle fits nicely against the cap and has six flutes about the circumference allowing you to mark three equally spaced locations.**

## The Mast

The 8-foot mast is made from two telescoping ¾- and two ⅝-diameter × 0.058-inch-wall aluminum-tubing sections. Slot the ends of the ¾-inch tubes so that they can be tightened around the smaller tubes with hose clamps.[5] To insulate the bottom ¾-inch section from the ⅜-inch bolt in the tripod that sup-ports the mast, its lower end is equipped with a plastic insulator. As shown in Figure 4, the insulator is a 2-inch length of acrylic tubing. The lower end of the acrylic tube extends about a quarter inch below the aluminum tube and is slotted so that the mast can be tightened around the bolt. Drill a 5/32-inch hole through the upper end of this insulator and the aluminum tube to pass a #6-32 bolt and nut to hold the insulator in place.

After mounting the SO-239 coax connector on the aluminum angle strip, solder a 2-inch length of #14 bare solid copper wire to the connector's center terminal and bend it close to the ⅜-inch bolt in the tripod top. Then bend the wire up and parallel to the bolt and about a quarter inch from it. When the bottom section of the mast is placed over the bolt, place this wire between the aluminum tube and a hose clamp. As you tighten the clamp, it makes the electrical connection from the coax to the mast and squeezes the slotted aluminum tube and insulator tightly against the bolt.

With the mast on the tripod, an easy way to make frequency adjustments is to separate the mast from its bottom section and lower it to the ground. You can then reach the flattop and coil without tilting the mast. For this reason, I don't tighten this joint. I place a #6-32 bolt through the ⅝-inch tube which is the second section of the mast, about one inch from its lower end so that it doesn't slide very far in. I still use a hose clamp over the ¾-inch tube, adjusting it to make a snug sliding fit for the upper mast.

For the top antenna sections, I use ⅝-inch-diameter thin-walled aluminum tubing used for aluminum clothes poles. This material is lighter and cheaper than the 0.058-inch-wall tubing used for the tripod and mast, but is strong enough. Short tubing sections can be joined together using 2-inch-long sleeves made from the ¾-inch-diameter × 0.058-inch-wall aluminum tubing. You need two 2-foot, two 1½-foot and two 1-foot lengths of the ⅝-inch thin-walled tubing, three couplings and a **T** joint to connect the flattop to the mast or the top of the coil. See Figure 5.

## Bungee Tie-Down

The antenna is quite light, and even with the wide base of the tripod it needs to be stabilized against wind gusts or someone tripping over the coax feed line. A bungee

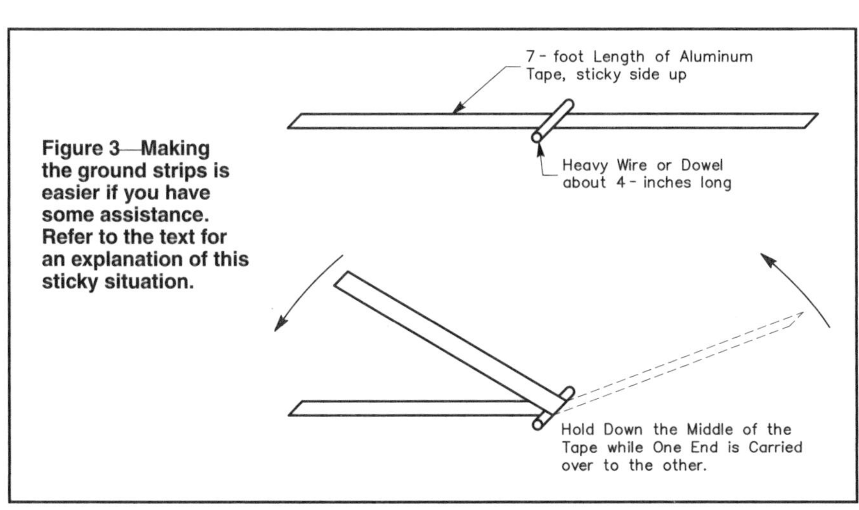


**Figure 3—Making the ground strips is easier if you have some assistance. Refer to the text for an explanation of this sticky situation.**

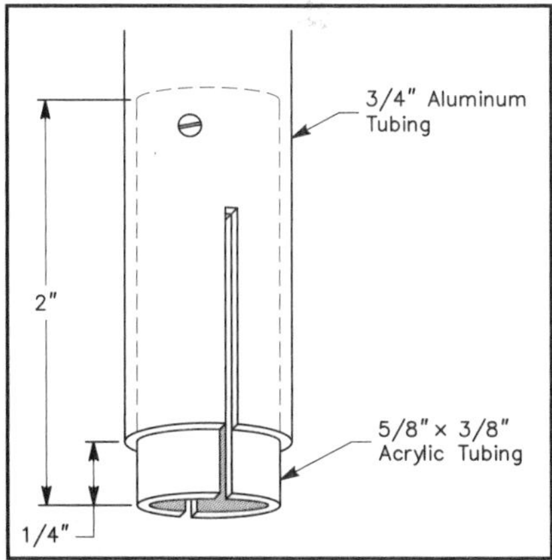


**Figure 4—An acrylic (Plexiglas) tube insulates the antenna mast from the ⅜×3-inch bolt that supports it. The tube has a ⅜-inch ID and ⅝-inch OD so that it slips over the supporting bolt and telescopes inside the lower ¾-inch mast section. Both the aluminum tube and the insulator tube are slotted using a hacksaw so they can be tightened around the bolt with a hose clamp. To mount a mobile antenna on the tripod, cut a 2-inch length of 1-inch-diameter acrylic rod and drill and tap one end to accept the ⅜×16 coarse-thread bolt of the tripod and ⅜×24 fine threads at the other end for the base of a mobile whip.**

cord and a ground stake[6] do an excellent job. The top of the tripod is about 3 feet high, so a 1/2- to 3/8-inch-diameter, 24-inch-long bungee cord works well. Any tent stake will do; drive it into the ground at an angle so it doesn't pull out easily. A special stake shaped like a large screw is ideal for this application.[6] It threads into the ground by hand and has a very low profile. (I leave the stake in the ground and my lawn mower doesn't even come close to striking it.) The stake won't go into hard, baked soil, however. For stability in such locales, or on pavement, hang some bricks, a rock or a jug of water beneath the tripod on the bungee cord.

## Antenna Operation on 10 through 17 Meters

For 10-meter operation, set up the tripod and place one end of a ground strip under each tripod foot. The ground strips may be laid in any direction. Adjust the mast to a length of about 7.4 feet. This length is quite a bit less than a quarter wavelength and I believe it's because of its closeness to the ground and the thickness of the tripod. No top hat is used on 10 meters. Adjust the mast length to resonate the antenna at your desired 10-meter frequency.

**Table 1**
**Length of thin-wall tubes needed for operation on 10 through 17 meters.**

| *Band (Meters)* | *Length of Flattop (ft) and Number of Sections* | *Length of Vertical Top (ft)* |
|---|---|---|
| 10 | 0 | 0 |
| 12 | 1 × 2 | 1.5 |
| 15 | 2.5 × 2 | 3.5 |
| 17 | 3.5 × 2 | 5.5 |

Table 1 provides lengths for the thin-wall tubes that you add to the antenna, either as a flattop or a vertical, for operation on 12, 15 or 17 meters. No change to the ground system is needed when changing bands. Table 1 assumes that you will leave the mast set for 10-meter operation. This simplifies band changing, such as moving from 10 meters to 15 meters and returning to 10 meters. These changes are quickly made by just adding the tubing lengths for 15 meters and removing them to return to 10 meters—no measurements, no tools.

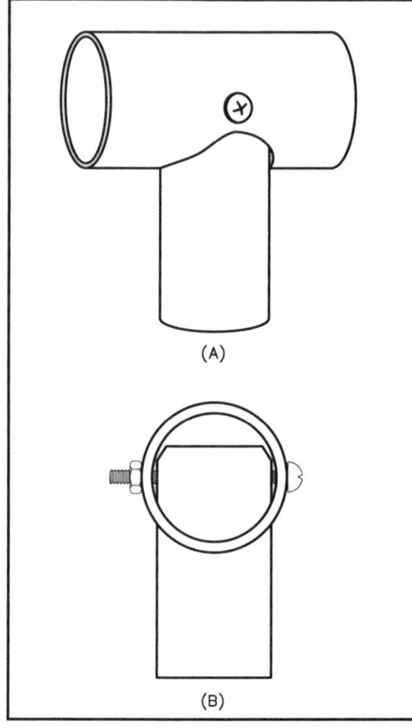


**Figure 5—A T is needed to make the flattop. The 3/4-inch-diameter horizontal tubing has a 0.058-inch wall and accepts the 5/8-inch-diameter thin-wall tubes. The 5/8-inch-diameter vertical piece has a 0.058-inch wall and fits into the 3/4-inch tube at the top of the coil form. To make a 5/8-inch hole in the 3/4-inch tube, drill a hole then expand it with a 5/8-inch-diameter or larger countersink. (This process is heavy work for a countersink, so use a little lubricating oil.) Before drilling a 7/64-inch hole for a #6-32 bolt through the T, assemble the two pieces and squeeze them together tightly in a vise, making them perpendicular. To mount a flattop on the mast without the coil, first place a 3/4-inch-diameter coupling sleeve over the 5/8-inch-diameter top of the mast and fit the T into that coupling.**

## 6-Meter Antenna Operation

For 6-meter operation, the tripod must be insulated from ground and the mast reduced to a length of 52 inches from tripod to tip; see Figure 6. No ground-coupling strips are needed. Simple insulators can be made from 1/2-inch CPVC pipe and couplings. Cut three lengths of pipe about 4 to 6 inches long and hammer each into a coupling. Cementing them isn't necessary; they will be a tight fit. The other side of the coupling fits well over the 5/8-inch-diameter aluminum-tubing leg. Adding these insulators to the tripod resonates the antenna in the 6-meter band with good SWR. You can change the operating frequency by adjusting the length of the mast only—you don't need to adjust the size of the tripod.

## Building the Loading Coil

For operation on the 20, 30 and 40-meter bands, a loading coil must be added to the antenna. A large tapped coil is shown in Figure 7; it tunes the antenna to 20, 30, or 40 meters and permits you to tune to the higher-frequency bands without changing the lengths of the top hat. The coil has 13 turns of #8 aluminum wire wound on a 4-inch styrene pipe coupling.[7, 8] This coil form is secured to a 7-inch-long, 1/2-inch-diameter CPVC pipe using 1 1/4-inch-long, #6-32 brass or stainless steel machine screws and nuts. I like to reinforce the 1/2-inch pipe by hammering a 2-inch length of 1/2-inch wood dowel into each end. This allows me to tighten the nuts and bolts without flattening the pipe. These bolts also secure the ends of the 13-turn coil. Using a marking pen, I made black marks on the coil to identify the fifth and tenth turns. The marks serve to locate the proper tap points without having to count coil turns each time.

Inside the styrene coil form is a ridge. Use a chisel or file to remove about a 1-inch-long section of this ridge to allow the CPVC pipe to lie flat against the inside of the form. Drill 7/64-inch holes at the ends of the styrene coil form and through the 1/2-inch pipe, then bolt them together as shown in Figure 7. Take a 16-foot length of aluminum wire, bend a loop at one end of it, attach the loop to one of the bolts and wrap the form as neatly as possible with 13 turns of wire, without bends, spacing the turns to

**Figure 6—Here, the antenna is set up for use on 6 meters. The tripod construction remains the same, using legs approximately 4 feet long, but the mast has been shortened. No ground strips are needed and the legs are insulated from ground by 1/2-inch CPVC pipe extensions at their feet.**

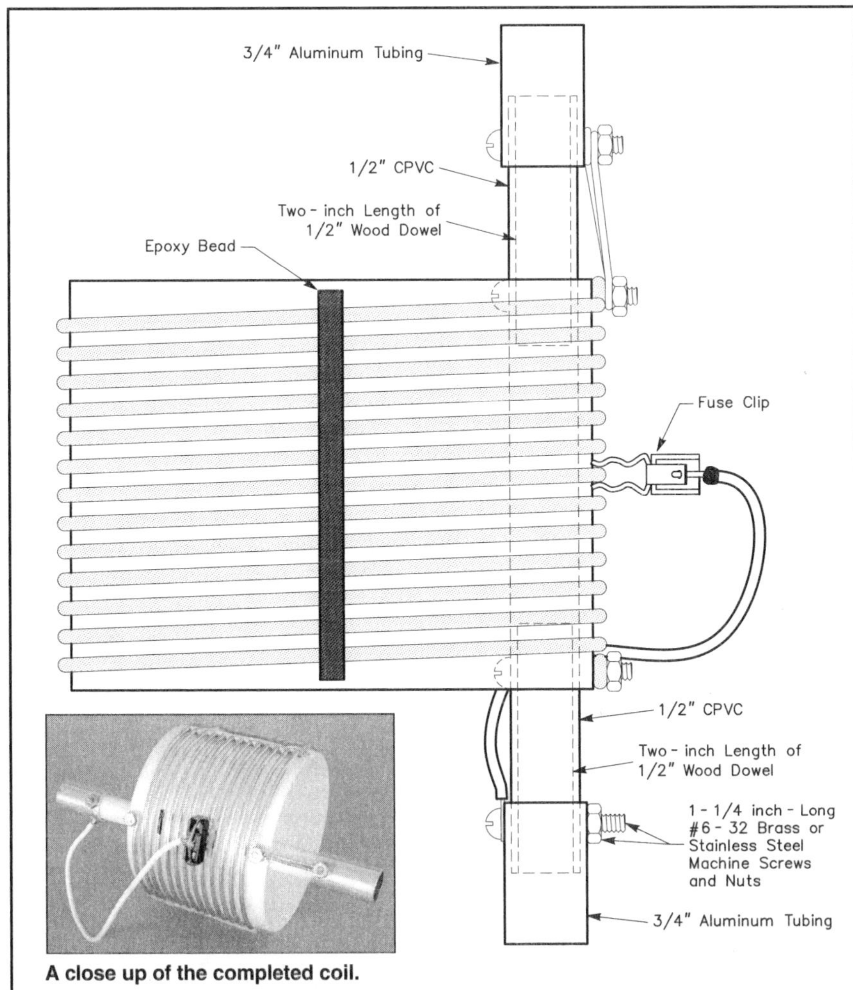


**A close up of the completed coil.**

**Figure 7—To make the loading coil, 13 turns of heavy aluminum wire are spaced to fill the form. Secure the coil ends using the same bolts that hold the plastic pipe inside the 4-inch styrene coil form. Mount the coil on the mast with a 3/4-inch-diameter aluminum sleeve at the bottom of the plastic pipe; the tap wire is also connected here. An identical sleeve at the top of this plastic pipe connects to the thin tubing for the top vertical section, or to an aluminum T to hold the flattop.**

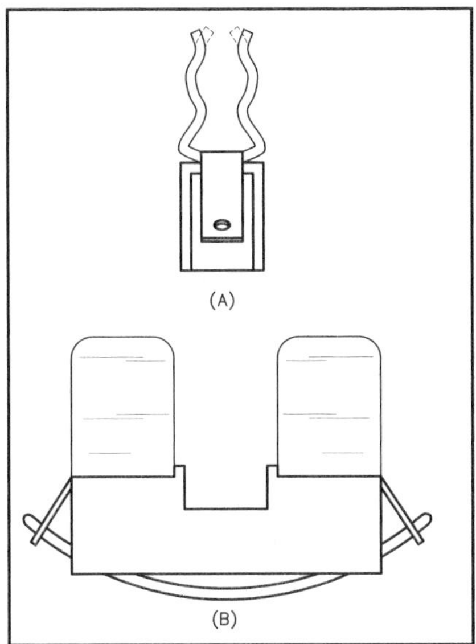


**Figure 8—Making the tap connection to the coil. At A, the ends of the jaws of a 5-mm cartridge-fuse holder are bent inward (dotted lines) to grip the heavy wire of the coil. A side view of the fuse holder is shown at B. Bend the solder lugs at the ends of the fuse holder to accept a wire passing through them and beneath the fuse-holder base. When this wire is in place, bend the lugs farther up against the ends of the holder and solder them. Strip a 1/4 inch of insulation from the tap wire and solder it to the wire joining the lugs beneath the fuse holder. Round off any sharp points or rough edges with a file, because you'll be gripping this connector tightly for attachment to and removal from the coil.**

fill the form. Wrap the end of the 13th turn around the bolt at the other end of the coil form and cut off the excess wire.

To tighten the wire on the form, clamp the form in a vise, grab the coil turns between both hands and progressively rotate the coil from one end to the other several times. This makes the turns tight enough to stay in place as you even out their spacing. To hold the turns in place permanently, run three ribs of epoxy the length of the coil. Use metal/concrete epoxy which has black resin and white hardener, making a dark gray mix that is easy to see against the white background of the coil form. One of these ribs is visible in Figure 7. To make nice straight ribs, first place strips of tape on each side of an intended rib location, apply the epoxy and remove the tape before the epoxy hardens.

Several types of alligator-clips will fit between the coil turns without touching neighboring turns, but I prefer to use a tap connection made from a fuse holder; see Figure 8.[9] After bending the fuse-holder-jaw tips, bend the jaws themselves to make them fit the wire tightly, but remain easy to attach and remove. Suit yourself as to how tight a grip they should have. Join the tap connector to the sleeve at the bottom of the coil form using a 9-inch length of stranded, insulated #14 copper wire, with a solder lug at the end. Use a similar piece of wire to join the top of the coil to the sleeve at the top of the coil form.

The sleeve at the coil bottom joins the coil to the mast. It is a 1 1/2-inch-long, 3/4-inch-diameter, 0.058-inch-wall aluminum-tubing piece. Insert the bottom of the 1/2-inch CPVC pipe halfway into this sleeve and drill a 7/64-inch hole through the sleeve and pipe. Fasten them together with a 1-inch-long, #6-32 brass or stainless steel machine screw and nut. The wire to the tap connector is attached with this same screw.

## Antenna Operation with the Coil

To use the antenna on 40 through 10 meters, shorten the mast to 6 feet 2 inches and connect the coil to the mast. Atop the coil, add an element consisting of two horizontal 3 1/2-foot lengths of 5/8-inch-diameter tubing, or a single 7-foot vertical piece of tubing. With the full 13 turns of the coil, and part of an extra turn supplied by the tap wire, the antenna will likely resonate in the middle of the 40-meter band. To operate at the low end of the band, add a 1-foot length of tubing to one side of the flattop or the vertical tubing section. See Figure 9 for approximate dimensions of the assembled antenna.

It may seem as though Table 2 has some errors because it lists a greater number of coil turns for operation on 15, 12 and 10 meters than for 17 meters! You're right—something strange is going on. It's because there are *two resonant frequencies* for each setting of the coil tap. Figure 10 shows the two paths that RF can take in the antenna. The upper part of the coil and the top hat provide the lower frequencies; the lower half of the coil provides the higher frequencies. A coil this large has considerable capacitance to free space, so it's not just an end-loading inductor at the higher fre-quencies. The antenna bandwidth is good, the SWR low and the antenna performs well on these bands. The charm of this coil system is that you can change

**Table 2**
This table identifies the number of coil turns (counted from the top of the coil) required to resonate the antenna on the 40- through 10-meter bands. These coil-tap settings are provided as a starting point only because installation conditions vary. To raise the antenna's operating frequency, reduce the number of turns used; to lower the operating frequency, increase the number of turns.

| *Band (Meters)* | *Number of Coil Turns* |
|---|---|
| 40 | 13 |
| 30 | 7.1 |
| 20 | 3.1 |
| 17 | 2 |
| 15 | 5 |
| 12 | 7 |
| 10 | 13 |

bands by just moving the tap on the coil, without any adjustments to the mast length or the flattop. And a bonus: With 13 turns on the coil, the antenna works on 40 and 10 meters simultaneously.

The coil settings of Table 2 may need some minor adjustments if a vertical top section is used instead of the flattop. In general, the SWR is lower with the flattop and the antenna is easier to handle.

## Power-Handling Capability and Safety

Because of the large coil and tubing used, you might be tempted to run high power with this antenna. I suggest you don't. The antenna may take it, but people can't. At high-power levels, dangerous RF voltages on the antenna are within range of physical contact. I have used the antenna at a 100-W level, but even that requires care and supervision.

## Other Possibilities

With the tapped coil, this antenna can be tuned to any frequency from 7 to 40 MHz when operated on the ground-coupled tripod, and up to 110 MHz with the tripod insulated from ground.

The antenna also may be used with a longer mast for greater efficiency, or with a shorter mast when space is restricted. Even though the short version is only about 6 feet high, you can't use it indoors because it must be coupled to earth ground. The taller antenna gets out better, but band changing is more complicated. If operation on 75 and/or 80 meters is a must, you can add another coil to the antenna just below the 40-meter coil and change antenna frequencies with the 40-meter tap. Adding a coil made of 20 close-wound turns of #12 enameled wire wound on a 4-inch styrene form similar to the one in Figure 7 will allow you to tune the antenna from about 3.5 to 3.8 MHz, and from about 3.8 to 4.1 MHz with the top hat reduced to one 3.5-foot section and one 2-foot section. Six ground-coupling strips will provide a lower SWR on 80. A small vertical like this is not very effective for short-skip ragchewing, however. A λ/4-wire draped over bushes, flower beds or low tree branches offers more high-angle radiation.

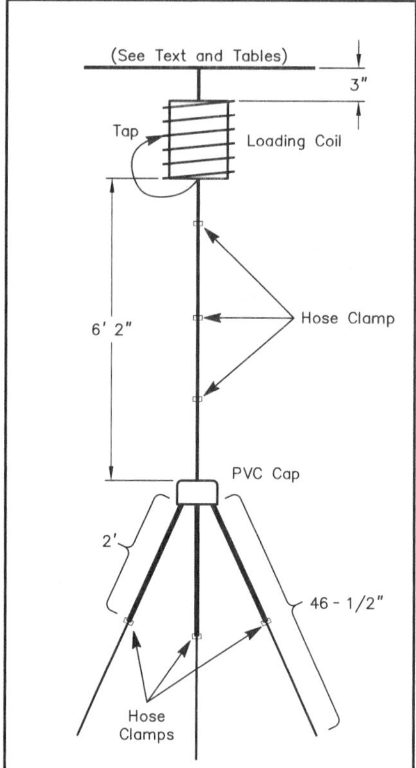


**Figure 9—Approximate dimensions of the assembled antenna with the tripod, mast, loading coil and top hat.**

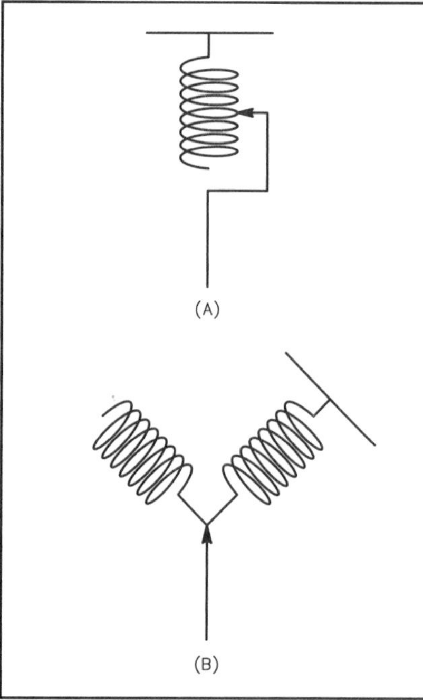


**Figure 10—At A, the upper part of the antenna includes the coil, the adjustable tap and the top hat. The bottom of the coil is free and not connected to anything else. At B, this has been redrawn to show the two antenna circuits with the two resonant frequencies that are present. The upper half of the coil has a lower resonant frequency because of the length of the top hat above it.**

### Notes

[1]Rick Lindquist, N1RL, and Steve Ford, WB8IMY, "Compact and Portable Antennas Roundup," 'Alpha Delta Outreach/Outpost System,' Product Review, *QST*, Mar 1998, pp 72-73.

[2]Twelve feet of each tubing size is needed. The aluminum tubing is available from Texas Towers and Metal and Cable Corp. See their ads elsewhere in this issue.

[3]The thin-walled 5/8-inch-diameter aluminum tubing is available from Home Depot and hardware stores as aluminum clothes poles, each about seven feet long.

[4]Adhesive-backed aluminum tape 2 1/2 inches wide is available from Home Depot stores in the heating-vent section.

[5]You may want to consider using an antioxidant at the tubing joints. Antioxidant compounds available from electrical wholesale supply houses, Home Depot and hardware stores include Noalox (Ideal Industries Inc, Becker Pl, Sycamore, IL 60178; tel 800-435-0705, 815-895-5181, fax 800-533-4483) and OX-GARD (GB Electrical, 6101 N Baker Rd, Milwaukee, WI 53209; tel 800-558-4311). Use either sparingly; a thin coat is sufficient.—*Ed.*

[6]Aluminum angle 1 1/2 × 1 1/2 × 1/16-inch thick is available from hardware and Home Depot stores. The green plastic ground stake that threads into the ground has the name "Twizelpeg" stamped into it, and is available at camping supply stores.

[7]The #8 aluminum wire is RadioShack #15-035.

[8]The coupling is available from Home Depot in the drainage pipe section, and also from large plumbing or swimming pool distributors. The couplings are actually 4 1/2 inches in diameter and made from polystyrene, a very low-loss insulator.

[9]RadioShack #270-738.

*Bob Johns, W3JIP, is an old gadgeteer who likes to play with antennas and coils. You can contact Bob at PO Box 662, Bryn Athyn, PA 19009;* **ksjohns@email.msn.com**.

*Photos by Joe Bottiglieri*

# An Off Center End Fed Dipole for Portable Operation on 40 to 6 Meters

## This compact design rolls up and fits in a small DX go-bag for easy travel.

**Kazimierz "Kai" Siwiak, KE4PT**

Your success with low power HF operations increases when you can increase the flexibility of your station. By that, I mean increasing the choices of operating modes, and choices of operating bands. The more choices you have, the more fun you will have. Here I describe an *off center end fed* (OCEF) dipole of my own design that is *physically* end-fed, but *electrically* off-center fed. I needed a simple-to-deploy antenna for portable use that I could easily stow in a small DX go-bag along with a Yaesu FT-817 radio and a tiny Elecraft T1 automatic antenna tuning unit (ATU). This OCEF dipole and ATU combination allows me to operate on all ham bands from 40 to 6 meters. I've even used it on the 80 meter band, with limited success. You can hang the OCEF dipole by one distant end support, or you can drape it in ad-hoc fashion over a fiberglass tent pole (Figure 1) or other convenient support — the ATU adds further flexibility. This lightweight antenna can be rolled and stowed in a small plastic bag for travel. Here's how I built mine.

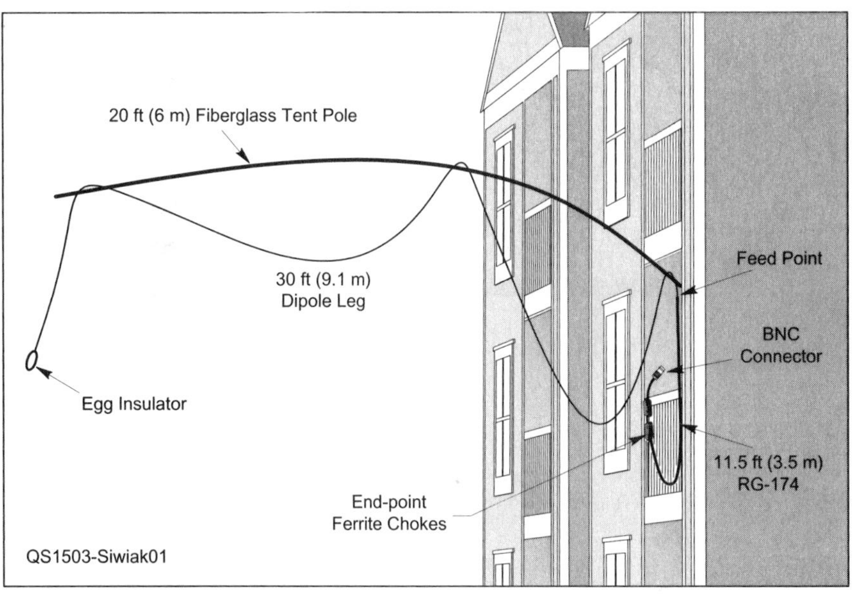


**Figure 1** — The OCEF dipole can be drooped from a fiberglass tent pole, as shown here, or stretched out and hung by the end egg insulator.

### OCEF Dipole Construction

The OCEF dipole consists of two dipole legs and an optional droop wire (Figure 2). For the far-end portion I used 30 feet (9.1 m) of Teflon® insulated #20 AWG stranded copper wire that starts at the ceramic egg insulator at A. You can also use PVC insulated wire. My wire was salvaged from a twisted pair with yellow and purple insulation, which I unraveled and joined together. Solder it to the center conductor of miniature RG-174 coaxial cable at the *electrical* feed point B. The second radiating portion is the outer shield of the 11.5 feet (3.5 m) long section of RG-174 coax that extends from the electrical feed point B to the ferrite chokes at the physical end of C.

Start with enough RG-174 coax (at least 13 feet; 4 m) so you can wind three turns of the RG-174 coax around each of two ferrite chokes, and terminate the coax in a BNC connector. I chose a BNC female connector so that I can easily add more feed line.

At point C, I added a 2-foot (60 cm) long Teflon covered #20 AWG stranded droop wire to help "tame" the typically high wave impedance at the end of a dipole. This optional wire C – D is the third radiating portion of the OCEF dipole. The electrical end of the antenna is C, or D (if the optional droop wire is used). The ferrite common mode chokes attenuate the radiating currents and keep them from flowing past the choke towards the transmitter.

### Theory of Operation

The OCEF dipole is 41.5 feet (12.6 m) long measured from A to C, or 42.4 feet (12.9 m) long from A to D. I chose the two dipole leg lengths A – B and B – C (or B – D) so that they would not be a multiple of a half wavelength on any band of operation.

The radiator length between the egg insulator and the end of the optional droop wire acts like a dipole radiator. This is a non self-resonant design, so you need an antenna match box — I use the Elecraft T1 automatic Antenna Tuning Unit (ATU) — to present a 50 Ω load to the transmitter.

You can attach the optional 2-foot (60 cm) long droop wire to move the very high wave impedance present at the end of a dipole away from the choke, to the end of the droop wire, easing the job of the ferrite chokes. If needed, you can wind the droop wire back around the antenna part of the coax to change the impedance seen by the ATU. Note that changing feed line lengths also changes the impedances presented to the ATU — I normally use a 15-foot (4.6 m) long cable between the antenna and the transmitter — you might need to change this length depending on how your tuner performs.

I chose the two dipole sections of different lengths so that the actual complex impedances at the dipole electrical feed point

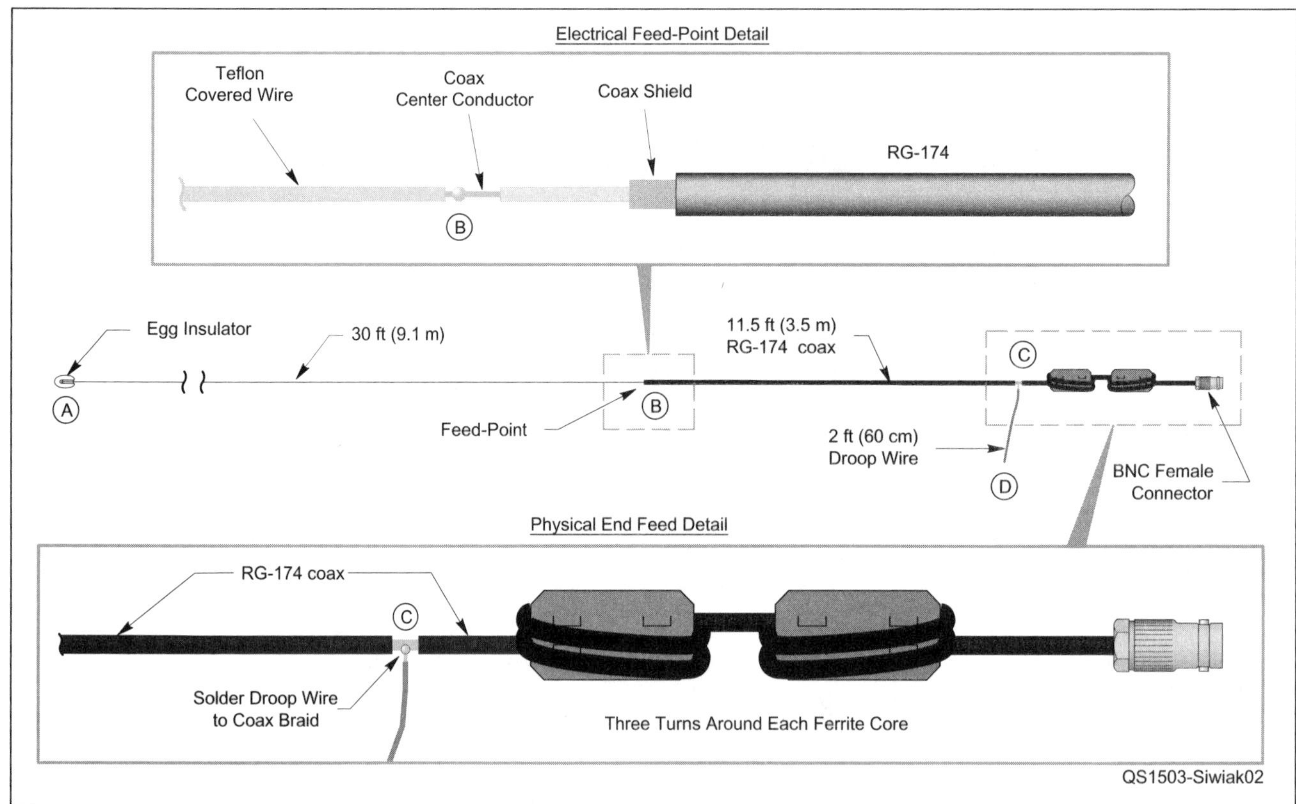


**Figure 2** — This OCEF dipole detail shows the radiating portions A – D, details of the electrical feed point, and the common mode chokes at the physical feed end. You need 32 feet (9.7 m) of #20 AWG stranded copper wire, Teflon or PVC insulated (part #125840 **www.jameco.com**), 13 feet (4 m) of RG-174 coaxial cable, 2 snap-on ¼ inch mix-31 split ferrites (**www.Palomar-Engineers.com**), and a cable-end BNC female connector.

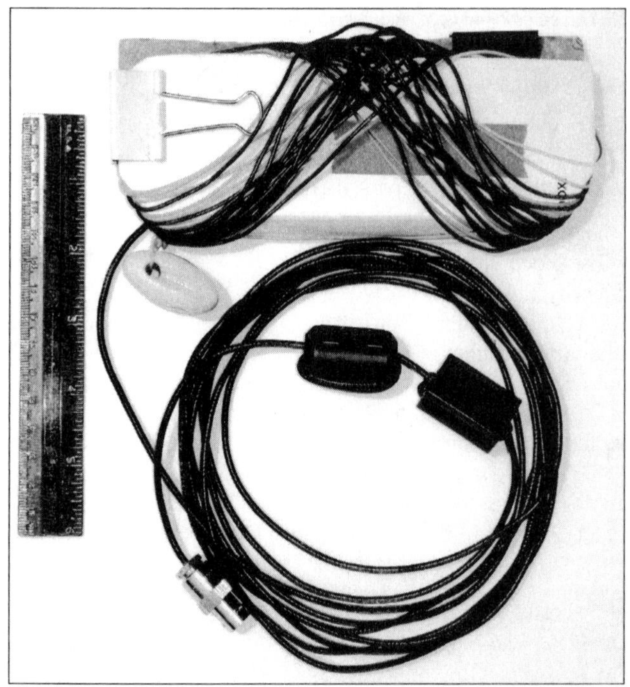

**Figure 3** — The antenna rolls neatly into a compact package for easy stowing (droop wire not shown). The ruler is 6 inches (30 cm) in length. [Kai Siwiak, KE4PT, photo]

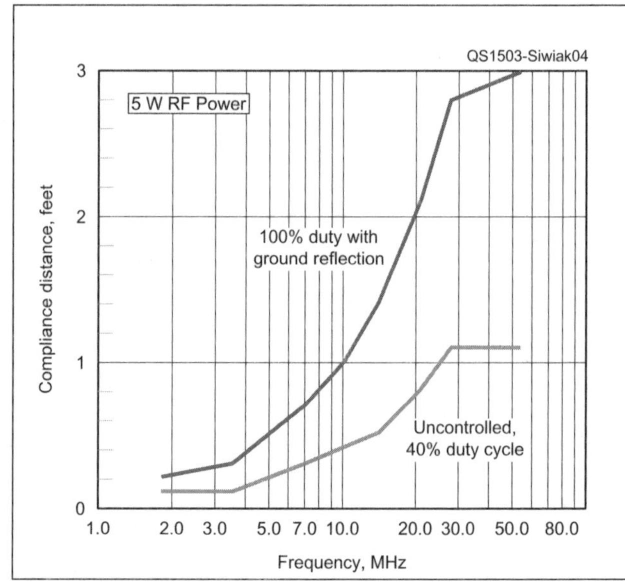


**Figure 4** — RF exposure compliance distance for 5 W power. For 50 W multiply the distances by 3.2.

would be relatively easy for the ATU to match. Again, no magic here, just a judicious choice of lengths.

### Deploying the OCEF Dipole

When stretched out and hung by the end insulator in an area clear of obstacles, the OCEF dipole is self-resonant near 11.8 MHz and its harmonics (slightly lower with the droop wire). We do not, however, use the antenna near any of the self-resonant frequencies. Antennas do not need to be self-resonant to radiate effectively. With this design, I can match the antenna using the ATU in any ham band from 40 to 6 meters — even and especially — if the antenna is deployed in a random fashion, such as drooped over a fiberglass tent pole, as in Figure 1. I've even had success on 80 meters when the antenna is fully stretched out.

### Stowing the OCEF Dipole

Starting at the egg insulator, I wind the wire section of the dipole in figure-eight style around a corrugated cardboard form (6 × 7 inches; 15 × 18 cm). Winding this way prevents twisting in the wire. I then coil the coax portion in a loop (Figure 3), and stow inside the folded cardboard form. The antenna easily fits in a quart (liter) size plastic bag along with additional connectors and jumper cables for the ATU and HF radio.

### Using the OCEF Dipole

With 5 W RF power the RF exposure compliance distance (Figure 4) from any part of the antenna rises steadily from 1 foot (30 cm) at 10 MHz to 3 feet (1 m) at 54 MHz. For 50 W RF multiply the distances by 3.2.

When packing constraints limit what I can bring along during travel, this antenna provides lots of freedom in choosing operating bands to let me get my share of DX. The antenna travels well and deploys easily from a hotel room balcony or from a tree at a camp site.

ARRL member Kazimierz "Kai" Siwiak, KE4PT, enjoys DXing and carries a low-power "DX go-bag" station while travelling. You can reach him by e-mail at **k.siwiak@ieee.org**.

**For updates to this article, see the *QST* Feedback page at www.arrl.org/feedback.**

By Rich Wadsworth, KF6QKI

# A Portable Twin-Lead 20-Meter Dipole

## With its relatively low loss and no need for a tuner, this resonant portable dipole for 14.060 MHz is perfect for portable QRP.

My first attempt at a portable dipole was using 20 AWG speaker wire, with the leads simply pulled apart for the length required for a $1/2$ wavelength top and the rest used for the feed line. The simplicity of no connections, no tuner and minimal bulk was compelling. And it worked (I made contacts)!

Jim Duffey's antenna presentation at the 1999 PacifiCon QRP Symposium made me rethink that. The loss in the feed line can be substantial, especially at the higher frequencies, if the choice in feed line is not made rationally. Since a dipole's standard height is a half wavelength, I calculated those losses for 33 feet of coaxial feed line at 14 MHz. RG-174 will lose about 1.5 dB in 33 feet, RG-58 about 0.5 dB, RG-8X about 0.4 dB. RG-8 is too bulky for portable use, but has about 0.25 dB loss. For comparison, *The ARRL Antenna Book* shows No. 18 AWG zip cord (similar to my speaker wire) to have about 3.8 dB loss per 100 feet at 14 MHz, or around 1.3 dB for that 33 feet length. Note that mini-coax or zip cord has about 1 dB more loss than RG-58. Are you willing to give up that much of your QRP power and your hearing ability? I decided to limit antenna losses in my system to a half dB, which means I draw the line at RG-58 or equivalent loss.

### TV Twin Lead

It is generally accepted that 300 ohm ribbon line has much less loss than RG-58. Some authors have stated that TV twin lead has similar loss as RG-58, which is acceptable to me. A coil of twin-lead is less bulky and lighter than the same length of RG-58. These qualities led me to experiment with it. One problem is that its 300 ohm impedance normally requires a tuner or 4:1 balun at the rig end.

But, since I want approximately a half wavelength of feed line anyway, I decided to experiment with the concept of making it an exact electrical half wavelength long. Any feed line will reflect the impedance of its load at points along the feed line that are multiples of a half wavelength. Since a dipole pitched as a flat-top or inverted V has an impedance of 50 to 70 ohms, a feed line that is an electrical half wave long will also measure 50 to 70 ohms at the transceiver end, eliminating the need for a tuner or 4:1 balun.

To determine the electrical length of a wire, you must adjust for the velocity factor (VF), the ratio of the speed of the signal in the wire compared to the speed of light in free space. For twin lead, it is 0.82. This means the signal will travel at 0.82 times the speed of light, so it will only go 82% as far in one cycle as one would normally compute using the formula 984/f(MHz). I put a 50 ohm dummy load on one end

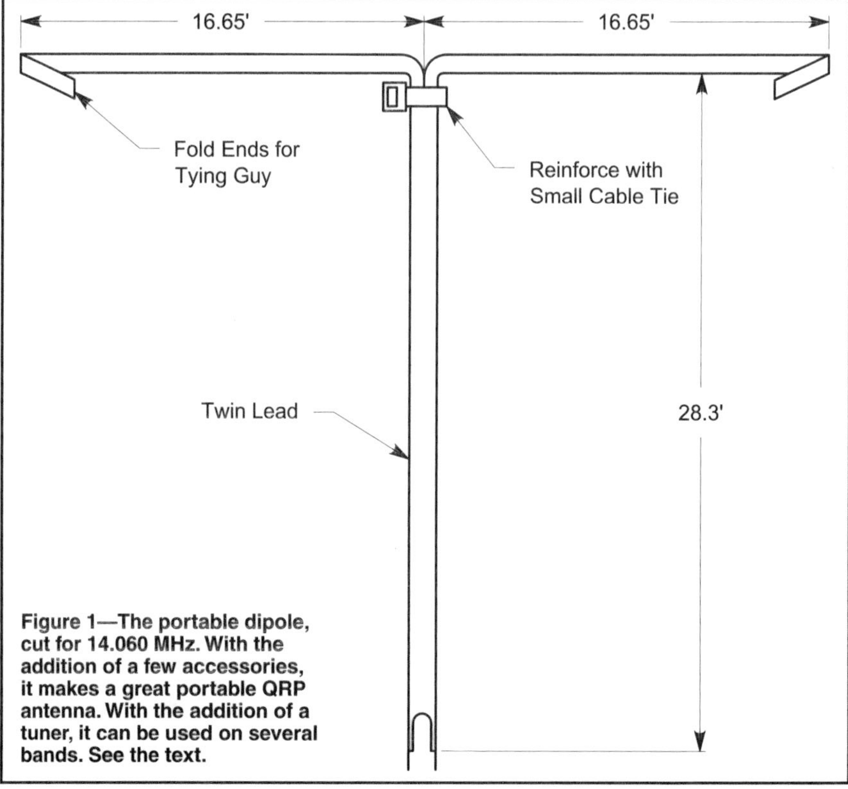


**Figure 1—The portable dipole, cut for 14.060 MHz. With the addition of a few accessories, it makes a great portable QRP antenna. With the addition of a tuner, it can be used on several bands. See the text.**

of a 49 ft length of twin lead and used an MFJ 259B antenna analyzer to measure the resonant frequency, which was 8.10 MHz. The 2:1 SWR bandwidth measured 7.76 to 8.47 MHz, or about 4.4% from 8.10 MHz.

The theoretical $^1/_2$ wavelength would be 492/8.1 MHz, or 60.7 feet, so the VF is 49/60.7=0.81, close to the 0.82 that is published. A $^1/_2$ wave for 14.06 MHz would therefore be 492×0.81/14.06 or 28.3 feet. I cut a piece that length, soldered a 51 ohm resistor between the leads at one end, and hoisted that end up in the air. I then measured the SWR with the 259B set for 14.060 MHz and found it to be 1:1. I used the above-measured 2:1 bandwidth variation of 4.4% to calculate that the feed line could vary in length between 27.1 and 29.5 feet for a 2:1 maximum SWR.

Now comes the fun part. With another length of twin lead, I cut the web between the wires, creating 17 ft legs, and left 28.3 feet of feed line. I hung it 30 feet high, tested, and trimmed the legs until the 259B measured 1:1 SWR. The leg length ended up at 16.75 feet. (Note: The VF determined above only applies to the feed line portion of the antenna.) There is no soldering and no special connections at the antenna feed point. I left the ends of the legs an inch longer to have something to tie to for hanging. I reinforced the antenna end of the uncut twin lead with a nylon pull tie, with another pull tie looped through it to tie a string to it for using as an inverted V. To connect the feed line to the transceiver, I used a binding post-BNC adaptor that is available from Ocean State.[1] My original intention of leaving the feed line free of a permanent connector was to allow connection to an Emtech ZM-2 balanced antenna binding post connectors. Since then I have permanently attached a short stub of RG-58 with a BNC, because I plan to either use it with my single band 20 meter Wilderness Radio SST, or with an Elecraft K1 or K2 with built-in tuner. I did this by connecting the shield to one side of the twin lead and the center conductor to the other side—no balun was used between the coax and twin lead.

After a year or so of use and further field testing, including different heights and V angles, I further trimmed the legs to a length of 16.65 feet. I found that the lowest SWR was usually obtained with the V as close to 90 degrees as I could determine visually. Also, I found that the resonant frequency (or at least the frequency at which SWR was at a minimum) is lower if the antenna is closer to the ground, and vice-versa. For example, with the top of the V at 22 feet, the lowest SWR was measured at around 13.9 MHz, and with the top of the V at 31 feet, SWR was lowest at around 14.1 MHz. In both cases, SWR at 14.060 did not exceed 1.3:1.

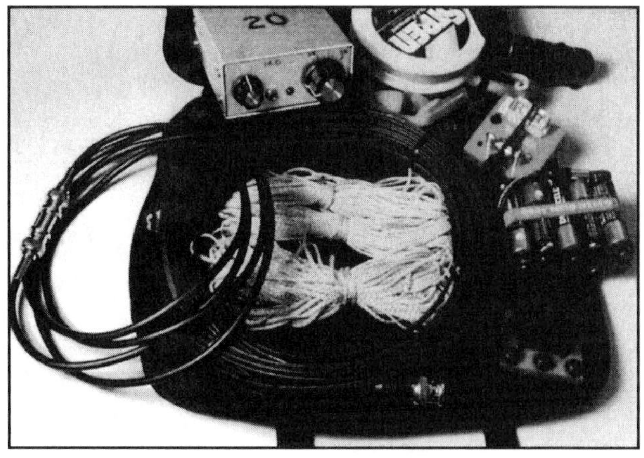


Figure 2—The author's portable station, including twin-lead dipole, 20-meter Wilderness Radio SST transceiver and support line. It all fits in the 8″×10$^1/_2$″×2″ Compaq notebook computer case.

I used Radio Shack 22 AWG twin lead that is available in 50 ft rolls. To have no solder connections, you need at least 45 feet. When I cut the twin lead to make the legs, I just cut the "web" down the middle and didn't try to cut it out from between the wires. It helps make the whole thing roll up into a coil, and the legs don't tangle when it's unrolled, since they're a little stiff. It turned out that the entire antenna is lighter than a 25 ft roll of RG-58. This antenna can be scaled up or down for other frequencies also. An even lower loss version can be made with 20 AWG 300 ohm "window" line, though the VF of that line is different and should be measured before construction.

## How High?

Wait, you say—"After all that talk about having it a half wave up, you only have it up 28 feet." A 6 or 12 ft RG-58 jumper, available with BNC connectors from RadioShack, can be used to get it higher if the right branch is available. Since impedance at the feed point is 50 to 70 ohms, 50 ohm coax can be used to extend the feed line. I have used it in the field a few times as an inverted V, at various heights and leg angles, and used an SWR meter to double-check its consistency in different situations. SWR never exceeded 1.5:1, so I feel safe leaving the tuner home. For backpacking, I leave the SWR meter home, too!

And there's a bonus: Since it has a balanced feed line, it *can* be used with little loss as a multi-band antenna, with a tuner, from 10 to 40 meters. I quote John Heyes, G3BDQ, from *Practical Wire Antennas*, page 18: "Even when the top of the doublet antenna is a quarter-wavelength long, the antenna will still be an effective radiator." Heyes used an antenna with a 30 ft top length about 25 ft off the ground on 40 meters and received consistently good reports from all Europe and even the USA (from England). It will not perform as well at 40 meters as at 20 meters, however, though 10 through 20 meters should be excellent.

## Testing, Testing

To test this theory, I recently worked some of Washington State's Salmon Run contesters and worked many Washington hams and an Ohio and a Texas station on 15 and 20 meters, with the antenna up 22 feet on a tripod-mounted SD20 fishing pole, using 10 W from an Elecraft K2 from central California. The K2 tuner was used to tune the antenna on 15 meters. Signal reports were from 549 to 599. Unfortunately, this was a daytime experiment and 40 meters was limited to local traffic.

At the 2001 Freeze Your Buns Off QRP contest, it was hung at 30 feet and compared to a 66 ft doublet up 50 feet on 10, 15 and 20 meters, using a K2 S-meter. There was little if any difference. At the 2001 Flight of the Bumblebees QRP contest I compared it, at 20 meters, to a resonant wire groundplane antenna with each antenna top at 20 feet and found it to consistently outperform the groundplane. I have concluded through these informal experiments that a resonant inverted V, when raised at a height close to or exceeding a half wavelength, produces the most "bang for your buck" and that extra length or height beyond that yields diminishing returns.

*A ham since 1998, Rich Wadsworth, KF6QKI, is a civil engineer in private practice as a consultant. Since earning his license, he reports, that he has become obsessed with kits and homebrewing. You can reach Rich at 320 Eureka Canyon Rd, Watsonville, CA 95076;* **richwads@compuserve.com**.

[1]Ocean State Electronics, 6 Industrial Dr, Westerly RI 02891, tel 800-866-6626 or 401-596-3080; fax 401-596-3590; e-mail: **ose@oselectronics.com**.

# A Paint-Pole Antenna

The old saying, "Where there's a will there's a way," certainly applies to hams who struggle with antenna restrictions. Living in a duplex town house complex, I was faced with the problem of putting up an adequate antenna for the HF bands. My first attempt was a dipole for 40 meters strung on the outside the walls of my unit in a horizontal **V** configuration. This was adequate to copy W1AW for code practice and make a few local contacts on CW. But when I upgraded to General, I wanted an antenna that would "get out" on the 20 and 17-meter bands.

I needed an antenna that could be assembled and disassembled quickly. Ideally, it wouldn't attract the attention of the neighbors or the landlord. I thought a vertical antenna would work, but the commercial verticals are not designed to be easily assembled and disassembled. As I browsed my local hardware store one afternoon, I literally bumped into the answer. There, propped in a corner, was a telescoping aluminum paintbrush extender—commonly called a "paint pole."

**Figure 1—The upper section of the paint-pole antenna (sections collapsed down). You can see the hose clamps and the copper braid I used to ensure electrical continuity between the sections. You can obtain copper braid by just removing it from an old piece of coax. Note the Radio Shack telescoping whip at the top of the antenna. It's held in place with electrical tape and a hose clamp.**

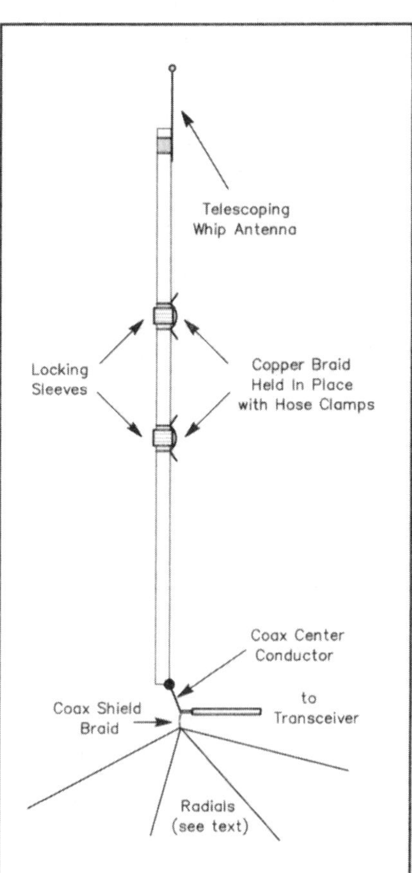


**Figure 2—Construction diagram of the paint-pole antenna.**

**Radio Shack *Superlock* fasteners (no. 64-2360)**
***Quickie* aluminum telescoping paintbrush extender (or equivalent): Quickie Manufacturing Company, PO Box 156, Cinnaminson, NJ 08007.**
**5 2-inch-diameter automobile hose clamps**
**Radio Shack replacement antenna (no. 270-1405)**
**⅛-inch sheet-metal screw and washer**
**2 4-inch pieces of copper braid**
**copper wire (for radials)**
**porcelain egg insulator (optional)**

### From Paint Pole to Antenna

The total length of the paint pole was 16 feet, which is about six inches too short for the middle of the 20-meter phone band. So, I bought a Radio Shack telescoping radio replacement antenna and clamped it to the end of the upper section of the pole. That gave me the extra length I needed. (It's a lot easier than it sounds!)

A small sheet-metal screw and washer were used to secure the center conductor of the coax at the bottom of the pole after part of the plastic handle was carved away with a hobby knife. I secured the pole to our deck fence with two pieces of Radio Shack *Superlock* Velcro fasteners.

A modest wire radial system was installed using copper wire (I used #17 wire, but almost anything will work). Part of the radial system is under the deck and part of it runs through a hedge. Four radials for 20 meters and four for 17 meters were cut to the appropriate ¼ wavelengths (16.5 feet for 20 meters and 13 feet for

The ingredients of your next HF antenna may be as close as your hardware store.

By Anthony J. Salvate, N1TKS
110 Prospect St
Greenwich, CT 06830

17 meters). The radials attach to the shield braid of the coax. I used a porcelain insulator as a junction point for my radials, but this is optional. A construction diagram and parts list are shown in Figure 2.

When the antenna was finished, I keyed my transmitter and measured the SWR on 20 meters. After making several adjustments to the *top pole section only*, I managed to achieve a 1.1:1 SWR. I marked the section at the locking sleeve so I'd be able to find the correct length again. I performed the same measurements on 17 meters and marked the top section accordingly. Fine length adjustments can be made by extending or collapsing the whip.

**But Does It Work?**

Now for the moment of truth. After listening to several conversations, I decided to call. I was ecstatic when I received a reply from a California station who gave me a 57 report. Another station in Arizona gave me a 59 report. I received a 55 from an Italian station on 17 meters, and a 59 on 17 meters from a station in Caracas, Venezuela! I had the biggest grin on my face when I heard the other hams chuckle at the story of my "paint-pole" antenna!

This antenna might not fit everyone's needs, but it's simple, effective and cheap (about $25). You can set it up and take it down in three or four minutes (Figure 3). I haven't tried the antenna on 10 or 15 meters, but it should work well, set to the appropriate length and with the appropriate radials. Have fun!

Figure 3—If you were a nonham neighbor, would this look suspicious? Probably not. You're likely to think it was a support for an umbrella. But in less than four minutes I can extend the antenna to its full height and get on the air. QST

## *Radio Tips*: DX QSL Managers

The easiest and usually most reliable way to get a DX QSL card is from a DX station's *manager*. Many active (and popular) DX stations use managers to lighten the load of incoming and outgoing QSLs. If DX stations are using managers, they'll usually tell you on the air. ("Please QSL to my manager, F6XYZ.")

A QSL manager is another amateur, often in a different country, who acts as a card clearinghouse. You simply send your QSL card to the manager, along with a self-addressed return envelope and an Airmail International Reply Coupon (IRC) or enough money to pay the return postage (usually $2). IRCs are available at your local post office. Some DX stations have QSL managers in the US. In this case, just send a self-addressed envelope with one unit of First Class postage. The manager checks your information against the DX station's logs and sends you the much-needed card as soon as possible.

When sending a QSL to a DX station's manager, there are a few tips that will make that volunteer's job easier. First, don't even bother to send a QSL (to a QSL manager or to any DX station) unless you show the time and date—in Universal Coordinated Time (UTC)—when the contact took place. Be sure that the self-addressed envelope that you supply is large enough to accommodate a normal-size QSL—unless you like folded cards!

Another technique that will help the manager is to write the call sign of the station that the QSL is for, as well as the date, time and band of the QSO on the back of the return envelope. That way, the manager can arrange the cards for easiest processing while he waits for the logs to come in from the DX station.

Which brings up another point. If you're slow in getting your card from the manager, don't harass him. The QSL manager is at the mercy of the mails, of the DX station and a number of other factors over which he has no control. He is a volunteer, doing a tough job. Treat him kindly. After a wait of some months, or if you have not received the card while everyone else in town have been showing off theirs for weeks, mail a polite note to the manager explaining the circumstances, and enclose an SASE for his reply.

# Fishing for DX with a Five Band Portable Antenna

## A telescopic fiberglass fishing pole supports five quarter-wave vertical monopoles for 10, 12, 15, 17, and 20 meters.

**Barry L. Strickland, AB4QL**

I enjoy getting out to the countryside with my portable rig and operating QRP on the five bands between 10 and 20 meters. While my Elecraft K1 was an easy choice, my antenna design evolved over time as the result of numerous experiments. The configuration I finally arrived at is five quarter-wave vertical monopoles in parallel and driven against 16 radials laid on the ground (see Figure 1). Having one tuned element per band means that I can switch between bands without the need for an antenna tuner. While a single portable vertical element is a trivial construction challenge, a portable antenna with five vertical elements proved to be a different case. First we'll look at the problems I had to overcome, as well as the mechanical aspects of this antenna, and then we'll look at the antenna's electrical characteristics.

### Design and Construction Challenges

The design challenges of this antenna were grouped into three categories: support of the antenna elements, prevention of coupling between the tuned elements, and construction techniques to enable its rapid setup and dismantling. Figure 2 summarizes the various mechanical aspects of the antenna that will be discussed below.

### *Support*

To support the antenna, I use a Shakespeare® Wonderpole fishing rod (model TSP16). The Wonderpole is a specialty pole designed for crappie and bream fishing. It is made of fiberglass and has five telescopic sections that extend from 45 inches in the collapsed state to 16 feet for this model. There is a wide variety of fishing poles available, ranging from stiff, short, and heavy surf rods, to lightweight and flexible fly rods. A crappie pole is a bit heavier than a fly rod and stiffer in the "spine," which makes it more suited to our use for antenna support. The Wonderpole is very light,

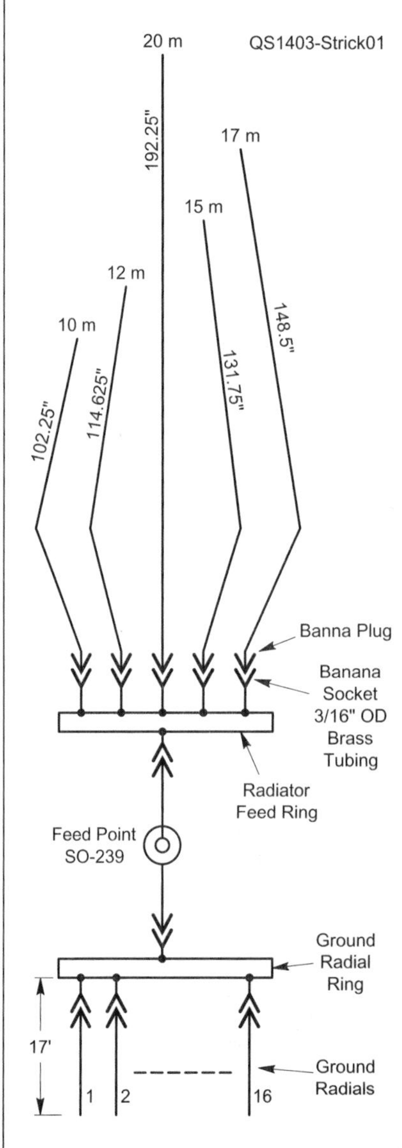


**Figure 1** — Schematic of the five band portable vertical antenna. The element lengths are those arrived at after tuning. In general, begin tuning with elements that are several inches too long and shorten them gradually in repeated round robin fashion (due to inter-element coupling) to obtain the minimum SWR for the band portion of interest.

which makes it ideal for portable operation, but it requires some additional support to keep it upright.

If the ground is not too rocky at the location where I want to set up, I drive either a wood or metal stake into the ground, taking care to keep it vertical. I then slip a length of ¾ inch PVC pipe over the stake and then slide the pole over the PVC pipe. The PVC pipe should slide about 10 inches into the pole and form a tight friction fit. If either the stake or PCV fits are too loose, they can be shimmed with tape for a snug fit.

As a backup for rocky ground or an extremely windy day, I made a three point guy ring (see Figure 3A). The material is not critical — you can use ⅛ inch acrylic or Lexan plastic or even a piece of scrap circuit board. I attached 10 foot string guy lines to anchor points on the ring using miniature carabiner clips similar to those used on key rings. The guy lines are tied to plastic bags filled with a few pounds of rocks collected at the operating site.

### *Coupling Prevention and Radiator Support*

The problem I faced with the five resonant quarter-wave radiators was how to support them more or less vertically while minimizing the coupling among them. My solution was to run the longest element — the 20 meter radiator — up the pole, letting it spiral around the pole a half dozen times before reaching the top. The other four radiators are held away from the pole by 18 inch spreader arms spaced at 90° intervals and emanating from a PVC hub that slides onto the Wonderpole from the top and stops about 44 inches from the bottom.

I made the spreader arms from ¼ inch diameter fiberglass rods that are sold in home improvement stores for use as reflective driveway markers. The antenna wire is secured at the end of each arm with a wire guide made from ¼ inch inside diameter

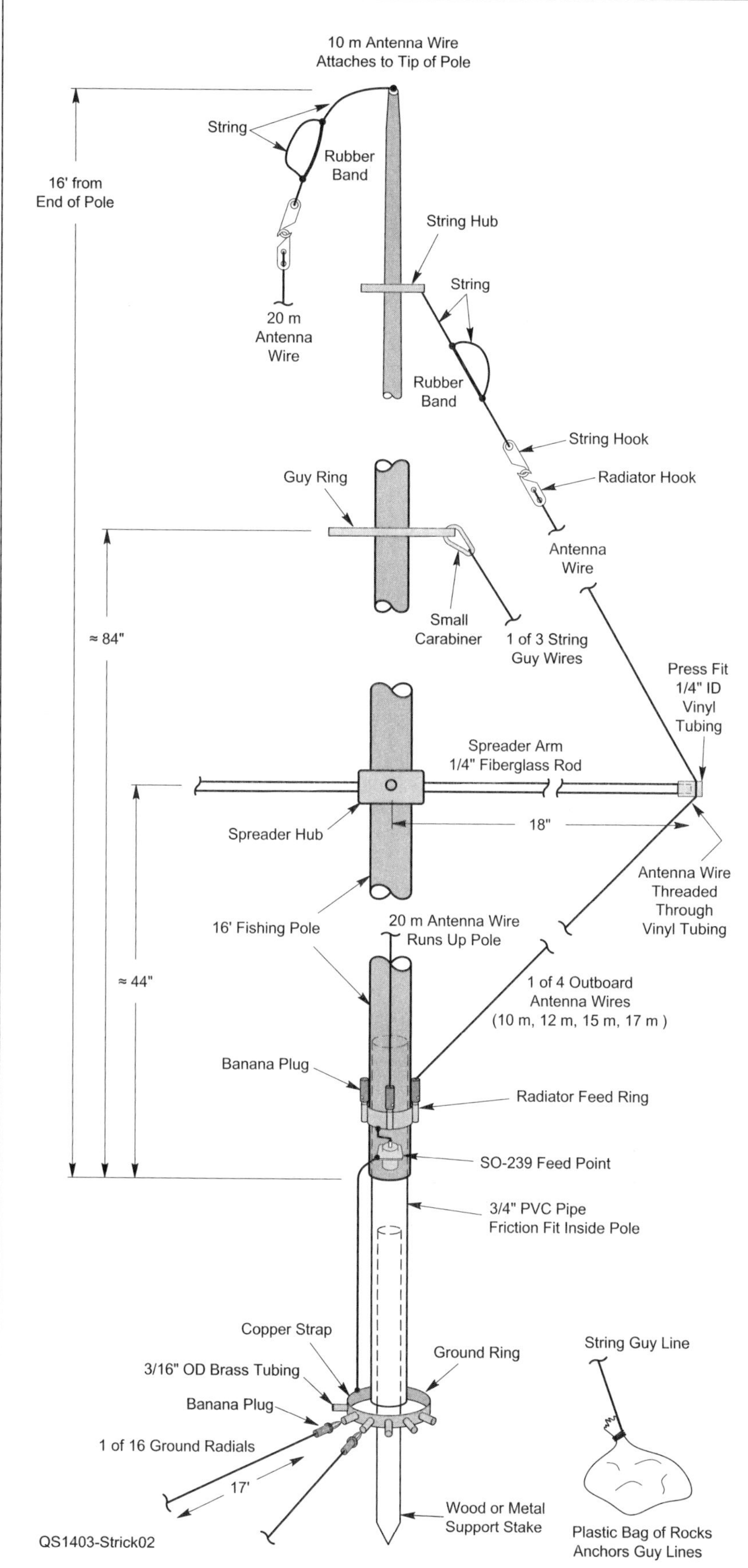


**Figure 2** — Mechanical details of the five band vertical antenna.

vinyl tubing that pressed onto the end of the spreader arm. Holes big enough to let the antenna wire (#20 AWG stranded insulated) slide through easily are drilled in the ends of the guides. The wire guides remain on the antenna wire for storage after dismantling.

While I was at the home improvement store looking for spreader arm material, I found the ideal item for the spreader hub — a ¾ × 2 inch PVC bushing. The spreader arms are inserted into four ¼ inch diameter holes drilled just under the flat part of the bushing (see Figure 4). Note that there are two walls to drill through for each hole. There is an inner shoulder on the ¾ inch bushing that should be filed away with a round file so that the hub can slide down the pole until it is about 44 inches from the bottom end. If you want, some of the excess hub material can be cut off. However, if you choose to use a power tool like a band saw, be extremely careful and use a jig or clamp to hold the work; do not use your fingers.

The four shorter radiators are supported by strings attached to a string hub that slides onto the pole and is held in place by friction (see Figure 3B). There is a 17⁄64 inch diameter hole in the center of the hub that limits its travel down the pole. At the free end of each string, a string hook (see Figure 3C) engages a radiator hook (see Figure 3D) that is attached to the end of each radiator. The radiator hook is held to the radiator line by the friction of the line passing through two holes in the hook. A small amount of tension is applied to each radiator by a rubber band tied into its supporting string. Thin plastic is adequate for the hooks — I cut mine from old credit cards.

The 20 meter radiator, which runs up the pole, is secured by a similar string and hook (see Figure 3E) assembly that is tied to the small ring at the tip of the pole.

At the bottom of each radiator is a banana plug that plugs into one of five sockets soldered to a radiator feed ring that is clamped to the bottom of the pole (see Figure 5). The sockets are fashioned from ¾ inch lengths of 3⁄16 inch outside diameter brass tubing commonly sold at hobby shops. It can be easily cut by rolling it back and

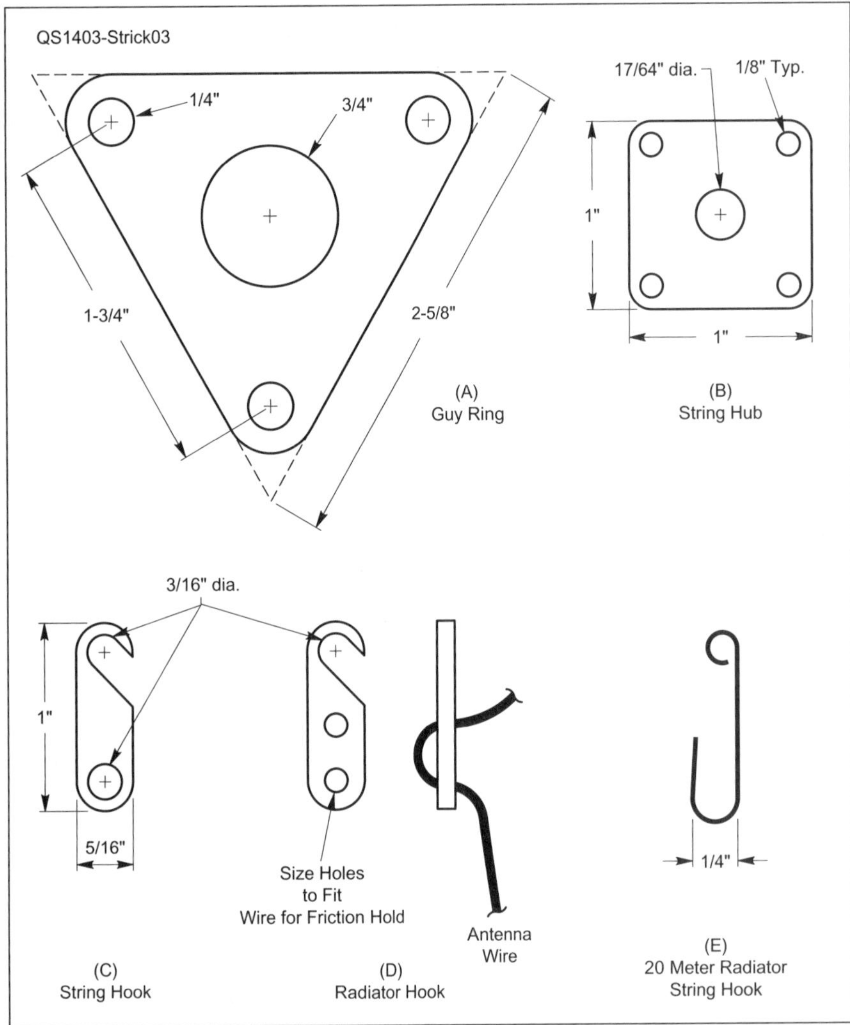


**Figure 3** — Detailed drawings of pieces used for supporting the pole and radiators: at (A) optional guy ring for pole when setting up over rocky soil or during windy conditions; at (B) string hub to support the four outboard radiators; at (C) radiator hook that holds wire fed through pair of holes; at (D) string hook that engages radiator hook to support outboard radiator wire; at (E) 20 meter radiator string hook is fashioned from paper clip in order to pass through string hub center hole.

**Figure 4** — Drilling hole through PVC bushing to form the hub for the spreader arms. Excess material on the spreader hub was later cut off to reduce its size.

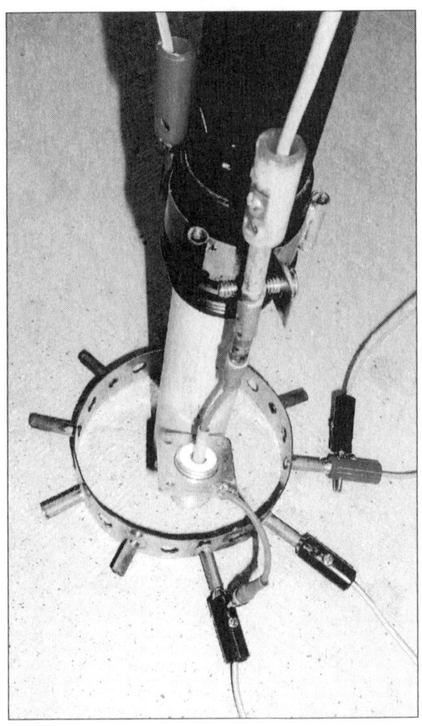

**Figure 5** — Bottom of antenna showing radiator feed ring clamped to end of five section telescopic fiberglass fishing pole that slides onto ¾ inch PVC pipe sleeve that slips over supporting stake driven into ground. The center conductor of the SO-239 feed point plugs into the bottom of the feed ring and the shell is connected to the ground ring. Connections are doubled up by using banana plugs with an auxiliary socket. Ground and feeder ring banana sockets are shop-fashioned from 3/16 inch OD brass tubing.

forth under a sharp knife. The feed ring is fashioned from 0.016 × 0.5 inch brass sheet typically sold alongside the brass tubing used for the sockets. One of the sockets is made double length to accommodate the banana plug from the center conductor of the SO-239 feed point.

The shell of the SO-239 feed point is connected via a short wire to a banana plug that connects to a ground ring made from a 3¼ inch diameter loop of copper plumbing strap with 10 brass tubing banana sockets soldered into existing holes in the strap that were enlarged to 3/16 inches to fit the brass tubing (see Figure 5). Although there are only 10 sockets, the ground ring can accommodate the 16 ground radials and feed point ground by using the auxiliary socket on the side of the banana jacks, which permits connections to be doubled up.

### *Rapid Setup*

The antenna sets up without the need for tools — all electrical connections are made with banana plugs and all mechanical connections use hooks or friction fit. The complete package is shown in Figure 6.

To deploy the antenna, I first extend the telescopic sections of the Wonderpole, which are held by friction. Adding a slight twist as they lock into place makes them extra secure. Next, I slide the radiator feed ring, spreader hub, guy ring (if needed), and string hub onto the pole. I then insert the four spreader arms into holes on the spreader hub. Now the radiator wires can be connected.

First I plug the 20 meter radiator, which is the longest, into one of the banana jacks on the radiator feed ring and route the wire up the pole with five or six evenly spaced turns around the pole. At the top, I hook the radiator hook, which is at the end of the wire, onto the string hook, whose attachment string is already tied to the eyelet at the tip of the pole.

Next, I install the four outboard radiators by first plugging each one into the radiator feed ring, then slipping its captive wire guide over the tip of a spreader arm and finally hooking its radiator end hook to the appropriate string hook, whose attachment string is tied to the string hub. Each string hook and its respective radiator hook is color coded to speed the setup (note that the upper face of the string hub is also marked so that the setup order is always the same). Before erecting the pole, I lay out the ground radials.

The 16 ground radials are deployed as follows: first, I pick a spot to erect the antenna and then drive the support stake into the ground. I then slide the ¾ inch PVC pipe sleeve over the support stake and slide the ground ring over the pipe sleeve. The ground radials are individually coiled for storage and secured with a hook-and-loop fastening strap. I remove the hook-and-loop fastening and throw the coil out from the antenna mount and then connect its banana plug to the ground ring. Afterwards, I pull all the radials taut and arrange them in an evenly spaced radial pattern.

Finally, I pick up the pole, slide it over the pipe sleeve and then connect the center conductor of the SO-239 to the radiator feed ring and its shell to the ground ring. All that's left is to connect my K1 to the SO-239 feed point and start calling CQ (see Figure 7).

## Electrical Aspects

In this section we'll look at how to tune the antenna, problems with modeling due to radials laid on the ground, some performance anecdotes, and finally a note on safety.

### *Tuning*

The design of the antenna was driven by the desire to minimize coupling among the five radiators. While the inter-element coupling was indeed reduced, it was not totally eliminated, which led me to adopt the following tuning method: First, find an area away from buildings and overhead power lines then set up the antenna and distribute the 16 ground radials in an even pattern. Next, attach an antenna analyzer to the feed point and begin tuning with the 10 meter radiator.

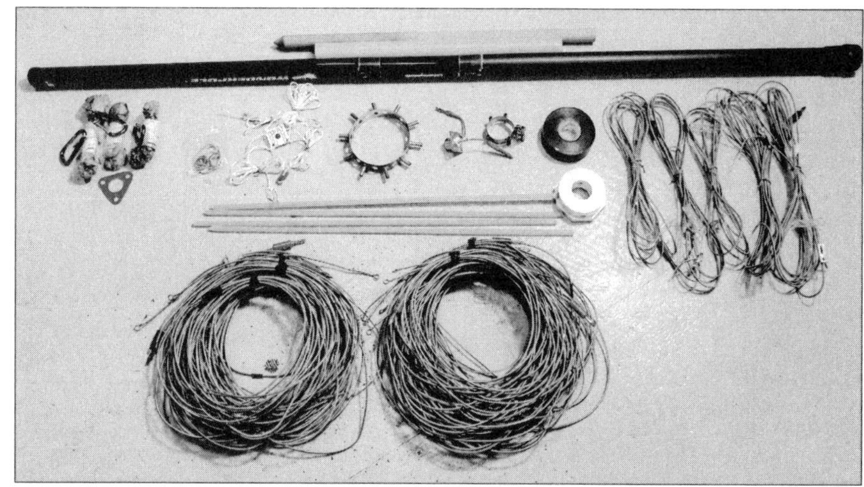

**Figure 6** — Entire antenna kit laid out. At the top: sharpened support stake that is driven into the ground; ¾ inch PVC pipe sleeve that slides over support stake; 16 foot, five-section telescopic Shakespeare® Wonderpole. Under the Wonderpole from left to right: backup red guy ring with three guy line sets comprised of miniature carabiner, 10 feet of guy string, and plastic bag to hold rocks to anchor guy string; bag of rubber bands; color-coded radiator support strings; copper ground ring; SO-239 feed point; radiator feed ring; electrical tape to back up tired Wonderpole friction joints; five radiator lines with captive wire guides on the four outboard lines. At the center and bottom: four spreader arms that fit into the white PVC spreader hub; 16 sets of ground radials, each secured with a hook-and-loop strap.

**Figure 7** — AB4QL operating CW QRP from one of his favorite remote locations in Mentone City Park on Lookout Mountain in northeastern Alabama. The antenna's low radiation angle and clear views of the horizon from sites like this have brought in numerous international QRP contacts.

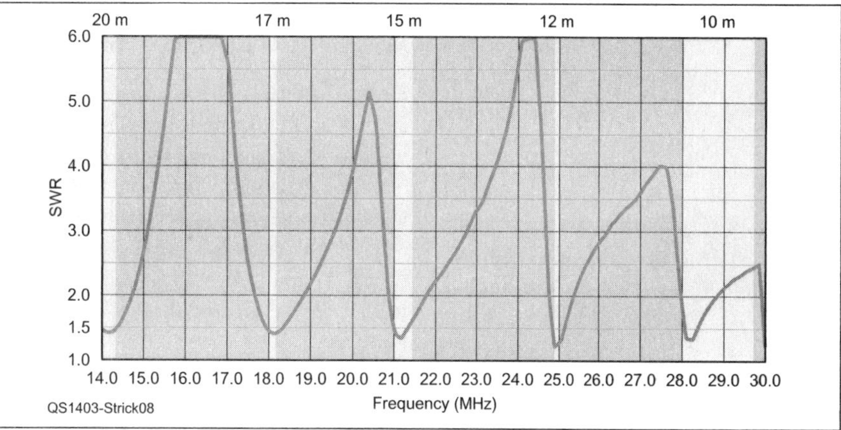


**Figure 8** — SWR plot from 14 to 30 MHz from RigExpert AA-54 antenna analyzer. Yellow highlighted regions are for amateur bands: 20 meters, 17 meters, 15 meters, 12 meters, and 10 meters respectively.

## Modeling a Ground Mounted Monopole Can be Tricky

The author mentions that he had difficulty modeling his ground mounted monopole to determine the radiation patterns. This is often a problem for *EZNEC* users, but it doesn't have to be, as we'll discuss. The first problem is that the usual amateur versions of *EZNEC* make use of the *NEC-2* calculating engine that does not allow modeling buried or on-ground radials. The professional version (*EZNEC PRO/4 V.5.0*) that does support the *NEC-4* calculating engine is quite expensive to start with ($650) and then it also requires a special license for the use of *NEC-4* ($300 for an academic or non-commercial license). This version supports the modeling of buried (but not on-ground or zero height) wires and has many other features, including the modeling of antennas with up to 20,000 segments as compared with 500 for *EZNEC V.5.0* ($89) or 1500 for *EZNEC+ V.5.0* ($139).[A]

### There are Some Solutions to This Problem

There is a simple technique, but as is often the case in real life, the results are not terribly satisfactory. By just modeling the monopole itself, with the bottom at zero height, we have a kind of model. If we select the GROUND TYPE (on the main *EZNEC* menu) as PERFECT, we get the elevation pattern shown in blue in Figure A. If we select the REAL/HIGH ACCURACY GROUND TYPE, we get the pattern shown in red. If we select REAL/MININEC GROUND TYPE, we get the pattern shown in black.

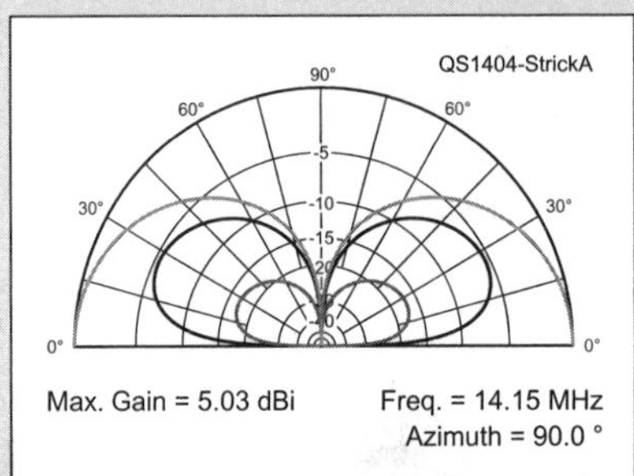


**Figure A** — *EZNEC V.5.0* elevation plots of the same 20 meter ¼ wave ground mounted monopole using the three different *EZNEC* ground models. The blue plot is with the *perfect* ground model, the red shows the results with the *real/high accuracy* model and the black shows the *real/MININEC* model results. For each of the real ground models, the conductivity was 0.005 S/M and the relative dielectric constant was 13. These are the default values representing "typical" ground, but can be changed if local information is available.

[A]The different versions of *EZNEC* antenna modeling software are available from developer Roy Lewallen, W7EL, at **www.eznec.com**. The models used in the sidebar are available on the *QST* in Depth web page at **www.arrl.org/qst-in-depth**.

As is clearly indicated, we can get whatever results we want by selecting the appropriate model — but which represents reality?

### What Do the Different Answers Really Mean?

My belief is that the *perfect* ground model is likely a reasonable representation of the performance of the monopole over seawater, an important case (see the contest results of Team Vertical as they operate from the water's edge of Pacific islands), but not one that most amateurs enjoy. We can get a calibration by running the author's 16 radial configuration, modeling the radials an inch or so underground using *EZNEC PRO*. The resulting elevation plot is shown in Figure B.

We note that the *EZNEC PRO* buried radial model elevation pattern looks strikingly similar to the pattern of the simple model over the *MININEC* ground model. This may be useful for an antenna with a substantial ground system such as this one, however, it does not indicate anything about the actual ground that is being used, a key element in the efficiency of antennas of this sort.

### A Solution is at Hand

Anyone who has an interest in this topic would do well to read about the excellent experimental work con-

Rather than making a single trim adjustment, only cut off about half the length needed to bring the radiator to resonance. Then move to the 12 meter radiator and perform the same partial trimming adjustment. Proceed in round robin fashion until all radiators show minimum SWR in the band portions that interest you. Remember, it's much easier to cut off a bit more wire than to add some back, so take your time. The SWR plot from 14 to 30 MHz made with my RigExpert AA-54 is shown in Figure 8.

### *Modeling*

Modeling with *EZNEC* proved to be problematic due to the radials that were laid on the ground. *QST* Contributing Editor Joel Hallas, W1ZR, was prevailed upon to explore this problem using the professional version of *EZNEC* and his findings are shown in the sidebar, "Modeling a Ground Mounted Monopole Can be Tricky."

### *Performance*

I'm aware that some folks may criticize a vertical antenna as noisy due to its omnidirectional pattern. This might be true in a suburban setting in proximity to plasma TVs, light dimmers, leaky power line transformers, and who knows what else in the way of electrical noise generators. However, away from cities, this is not the case, and in fact, the vertical's omnidirectional pattern gives me the freedom to operate in any direction without the need to worry about the fortuitous placement of trees that would enable me to string a horizontal antenna.

This antenna has been a good performer. Running CW QRP at 5 W or less from remote locations such as Mentone City Park on Lookout Mountain in northeastern Alabama, I have made contacts on all five bands with Europe, Asia, South and Central America, Hawaii, Alaska, and all over the continental United States. However, I'm still looking forward to the day when I receive a "G'day, mate!" from a VK down under. I'm sure it will come in time.

### *Safety*

My principal safety concern, operating at 5 W or less, is not to set up too close to a mountain precipice. However, others, perhaps using this antenna in the backyard, should be aware that on 10 meters if the

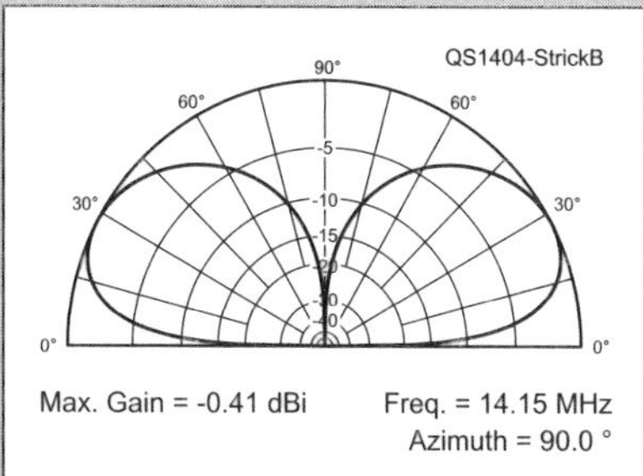


**Figure B** — *EZNEC PRO/4 V.5.0* elevation plot of the author's 20 meter ¼ wave ground mounted monopole with 16 radials (buried 0.1 feet deep). This analysis used the high accuracy ground model, the only one usable for buried wires.

ducted by Rudy Severns, N6LF.[B] One of Rudy's conclusions was that radials on the ground will act very much like radials under the ground. I took the buried radial model and changed all heights so that the radials were 0.05 feet above ground, eliminating the buried radial issue. Just as Rudy predicted, the results were virtually identical to those of the buried model using *NEC-4*. This was the case for each of the real ground models, and with both *NEC-2* and *NEC-4* calculating engines. In my opinion, this is probably the best *NEC-2* approximation of a buried ground system, since it takes into account the actual ground radial configuration.

[B]R. Severns, N6LF, "An Experimental Look at Ground Systems for HF Verticals," *QST*, Mar 2010, pp 30 – 33.

A simpler technique that I frequently use makes use of another result of Rudy's work. Rudy noted that four resonant elevated radials provided performance similar to up to 60 buried radials. Instead of modeling all 60 radials (a busy trig project to get all the coordinates), I just model an antenna with four ¼ wave elevated radials a foot or two off the ground. For this case, with the radials 1 foot above ground, the gain is exactly the same as that of the more complicated models. The additional height of the antenna does reduce the main elevation lobe by 1°, which is not a major discrepancy, in my opinion.

### Some Observations on the Design

In the process of performing these modeling experiments, I did note a few results worth mentioning, because they may affect the way the antenna works in the field. First, while radials right above ground tend to act like buried radials and are not terribly length sensitive, elevated radials do need to be resonant to work properly. Depending on the nature of the ground and how tightly the radials are secured, this antenna may work better if the radials are cut so that there are two opposite each other cut for each band. I would hedge my bets and leave four at 17 feet and then have the others cut to resonance as noted — especially if the radials are loosely laid on rocky ground. This change could be made in the field if the tuning is not what is expected.

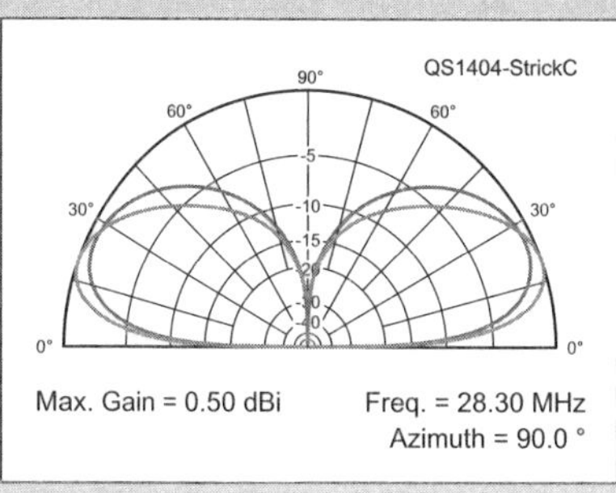


**Figure C** — *EZNEC V.5.0* elevation plots of the ¼ wave ground mounted monopole on 10 meters. The blue plot shows the pattern with the 20 meter wire in place and not connected. In that configuration, the SWR is about 5:1. The red plot shows the effect of removing, or detuning, the 20 meter wire returning the SWR to 1.5:1 with a normal ¼ wave pattern.

One more item I noted was interaction between the 10 meter monopole and the 20 meter wire, the only pair of harmonically related bands. If the 10 meter wire is driven, the ungrounded 20 meter wire — a ½ wave on 10 — actually improves the 10 meter pattern by increasing the gain and lowering the peak of the elevation lobe (see Figure C). The problem is that the Z goes to around 220 Ω, for about a 5:1 SWR. With a short length of coax and an antenna tuner, it will be a plus. To go back to the regular ¼ wave monopole pattern and SWR, fold some of the 20 meter wire back while using the 10 meter wire. — *Joel R. Hallas, W1ZR,* QST *Contributing Editor*

peak envelope power (PEP) to the antenna exceeds 50 W, the FCC requires that an RF environmental evaluation be performed. More about antenna evaluation can be found on the ARRL website and in Ed Hare's, W1RFI, book on RF exposure.[1,2]

I look forward to discussions with other amateurs who are building this antenna.

[1]**www.arrl.org/fcc-rf-exposure-regulations-the-station-evaluation**

[2]Ed Hare, W1RFI, *RF Exposure and You*, (Newington 1998, 2004). Available from your ARRL dealer, or from the ARRL Store, ARRL order no. 6621. Telephone toll-free in the US 888-277-5289, or 860-594-0355, fax 860-594-0303; **www.arrl.org/shop/**; **pubsales@arrl.org**.

Photos by the author.

Barry Strickland, AB4QL, holds an Amateur Extra class license and has been an ARRL member since he was first licensed as KA4LKH in 1979. Barry is a Volunteer Examiner and is active in the DeKalb County Amateur Radio Club. He has worked closely with the local Emergency Management Agency for a number of years as a Communication Specialist and was remotely deployed to provide communications after a 2012 tornado outbreak in northern Alabama. Now retired from the construction industry, Barry enjoys getting out to the countryside with his portable equipment and operating CW QRP. You can reach Barry at **ab4ql1@gmail.com**.

**For updates to this article, see the *QST* Feedback page at www.arrl.org/feedback.**

# A Super Duper Five Band Portable Antenna

*Make camping and travel more fun with this easy to duplicate antenna.*

**Clarke Cooper, K8BP**

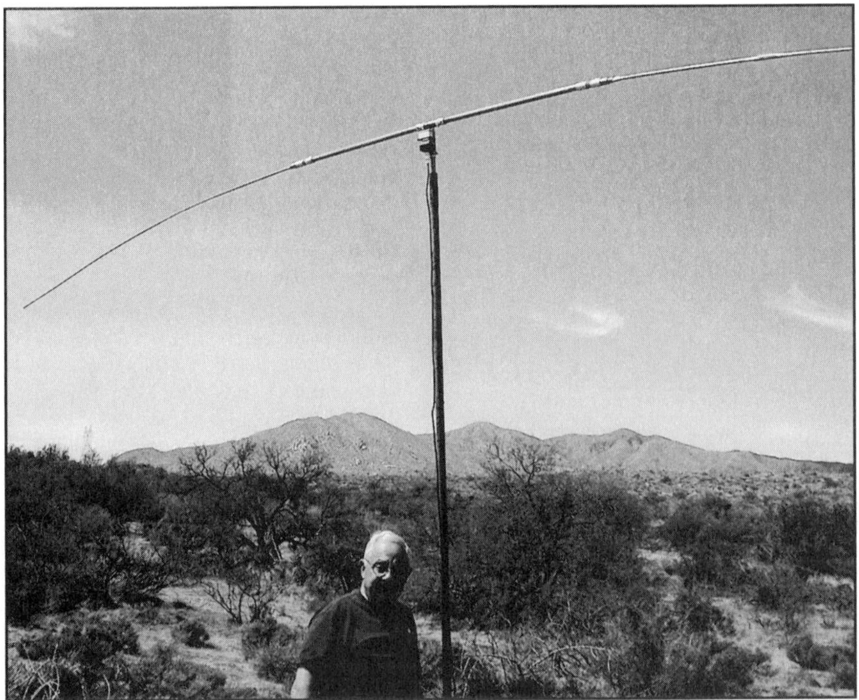

One thing that makes Amateur Radio so wonderful is whether I am camping in the rugged Grand Canyon area or somewhere in the Great Northern Woods, I have the capability to work the world while operating low power (QRP).

As one who loves building and experimenting with antennas and operating QRP, it seems like every time I venture out into the backcountry I bring a new homebrew antenna to try out. Then in March 2006 while I was looking through an issue of *QST*, I noticed an MFJ advertisement for their 10 and 12 foot telescoping whip antennas. That's when my wheels started churning. After a few minutes of doing some calculations and sketches on paper, I came up with this rugged lightweight antenna.

After placing the order with MFJ for two of the MFJ-1954 10 foot whips, I ventured to my local Home Depot and purchased the rest of the parts. Surprisingly, this little project only took me about an hour to build. The fun really started several days later when I received the telescoping whips. The initial tune-ups with my MFJ antenna analyzer and Elecraft K2 indicated that this antenna loads up beautifully on 20, 17, 15, 12 and 10 meters without the need for a tuner. The best news is that it can be built for around $60.

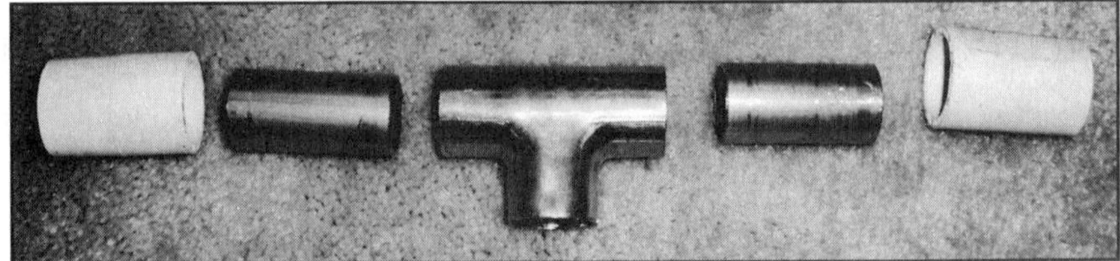

**Figure 1 — Details of PVC and copper pipe pieces for antenna center.**

## Travel Friendly

When broken down for transport, this antenna, minus the mast, consists of two short 2 foot 7 inch long elements and two lightweight MFJ whip sections retracted to a length of 2 feet.

To operate on the 20 and 17 meter bands, loading coils are used to resonate the system, with precise tuning available using the telescoping whips. In order to operate on the 15, 12 or 10 meter bands, the loading coil is bypassed and with a short jumper across the coil terminals the antenna operates as a full sized unloaded half-wave ($\lambda/2$) dipole.

## How Does It Perform?

Three days after completing my new antenna, I drove north from my winter residence in Phoenix to Grand Canyon National Park for a couple of days of camping. With this antenna 10 feet in the air and my Elecraft K2 at 5 W output, I had no trouble making SSB and CW contacts to all parts of the US as well as some long haul DX contacts. The DX contacts on that trip included Asiatic Russia, Australia, Belize, Lithuania, Italy, Japan and Hawaii on 20, 17 and 15 meters. My location was surrounded by high mountains to the north, east and west. I was quite pleased with my latest antenna's performance. Even though I found no activity at all on the 12 and 10 meter bands I still feel very confident that when they open up, this antenna will be a winner on these bands as well. I just can't wait until I start using this antenna on Pactor to get my e-mail.

## Making it Happen

The parts required are listed in Table 1. Cut the copper and PVC pipe per Table 1 and Figure 1. Remove burrs from all cut edges using emery cloth. Assemble all the parts on a clean surface. At a small sacrifice in weight, a ⅝ inch by 2 foot dowel can be inserted into the copper T to provide additional strength. This is recommended for extended use.

Push two of the copper bushings into the center T and insert the two CPVC couplings into the copper bushings as far as possible.

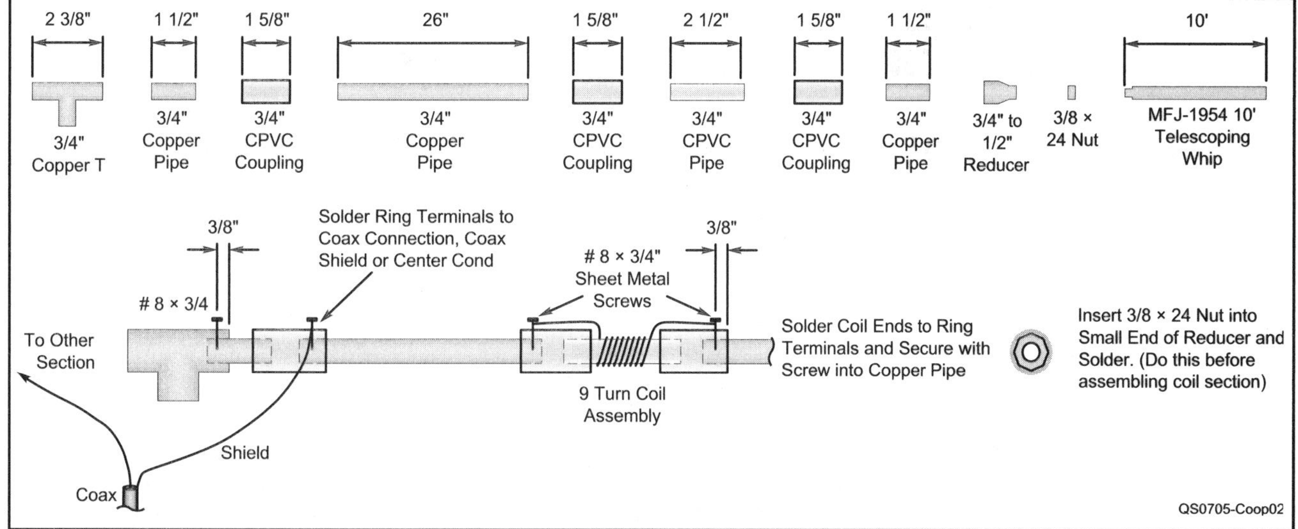


**Figure 2 — Antenna assembly drawing.**

**Table 1**
**Parts Required to Assemble the Portable Dipole**

*Quantity one unless noted.*

Alligator clips (4)
Copper pipe, 3/4" × 26 (2)
Copper pipe bushings (made from 3/4" pipe), 3/4" × 1½" (4)
Copper pipe reducer, 3/4" × ½" (2)
Copper T, 3/4"
CPVC cement
CPVC gold pipe, 3/4" × 2½"
Fine thread stainless nuts, 3/8" × 24 (2)
FSC hardwood dowel, 5/8" × 4'
Insulated solid 20 gauge wire, 10'
MFJ -1954 (10 foot) telescoping whip (2)
Sheet metal screws, #8 × 3/4 (package)
SXC CPVC couplings, 3/4"
Ring terminals, 20 gauge size (6)

**Table 2**
**Tools Required**

| | |
|---|---|
| Carpenter's saw | Hacksaw |
| Tubing cutter | Measuring tape |
| Electric drill and bits | Propane torch |
| Emery cloth | Soldering iron |
| Fine file | |

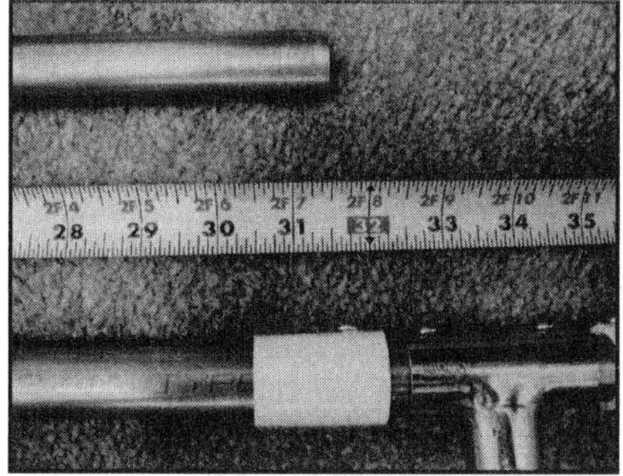

**Figure 3 — Detailed view of inner element construction.**

Drill four 3/32 inch holes on top of the center insulated T assembly and secure with four #8 × 3/4 inch screws.

Next, insert both 3/4 × 26 inch copper pipe elements into the CPVC couplings and drill four 3/32 inch holes. Then secure the four #8 × 3/4 inch screws as shown in Figures 2 and 3.

Assemble the loading coil forms by lightly gluing two CPVC couplings to each end of a 3/4 × 2½ inch CPVC pipe piece using CPVC cement. Make sure you apply glue only on the coupling half that will be used with the 3/4 × 2½ inch CPVC pipe, as the other end of the couplings will be fastened later by #8 sheet metal screws. After assembly, wipe any excess glue from the outside of the form and from the inside of the free end of both of the coil CPVC couplings.

Drill two vertical holes in both coil assemblies, all the way through each end of the CPVC coil form assembly. Insert a length of 20 gauge wire all the way through both vertical holes of the coil form. On the short end, leave enough wire to connect to a ring terminal later. Carefully wind nine turns on the form and feed the remaining end through both holes. Adjust the coil windings by twisting the coil and pressing on the coils to remove any slack. When you are satisfied that the coil windings are tight, leave enough wire to connect to a ring terminal later. The completed coil assemblies are shown in Figure 4.

Insert the remaining 3/4 × 1½ inch copper bushings into a 3/4 × 1½ inch copper pipe reducer and carefully solder them together. Handle with extreme care until these items have cooled down.

Insert both 3/4 × 24 nuts into the ½ inch ends of the pipe reducer until they are flush with the ends of the reducer. Carefully solder the nuts to the copper reducer. After they have cooled down, insert the 3/4 inch ends of the reducers into the 20 meter coil form. Confirm that both coil windings run in the same direction. Then drill a hole in each outer section of the CPVC bushing, keeping both holes symmetrical.

Measure the coil wire tag ends and carefully strip off just enough insulation to make a loop that feeds into the terminals. Confirm that the center of the terminal is aligned with the mounting hole. Crimp and solder the wire connections.

Fasten each coil form to its respective pipe reducer and secure with a #8 × 3/4 inch sheet metal screw. Slide the remaining end of each loading coil housing onto the end of each 2½ foot dipole element. Rotate the coil terminals so they are on top. Secure the coil assembly to each dipole element with a #8 × 3/4 inch sheet metal screw. Attach the MFJ -1954 telescoping whips by screwing them into the end of each coil assembly as shown in Figure 6.

Attach a length of coax by separating the center conductor and outer shield and soldering a ring terminal on each. Secure each terminal to the outer side screws on the T assembly. Wind a 6-turn 7 inch diameter coil on the antenna end of the coax and secure with tape or wire straps. Connect the coax to the antenna by loosening up the side screws on each active element insulator. These screws are the ones on the outer edge of the CPVC coupler next to the element. Slide the coax terminals under each of these screws and retighten the screws. It doesn't matter which side the coax shield or center conductor is attached to. After these connections have been made, take an ohmmeter and hold one meter lead on the center T and with the other lead touch both elements. Both sides should indicate an open circuit.

If you wish, instead of the coiled coax

Figure 4 — Completed loading coil assemblies.

**Table 3**
**Antenna Tuning Chart Settings and Resulting SWR**

| Band | Tune Frequency (MHz) | Max SWR | Coil | Whip Length |
|---|---|---|---|---|
| 20 Meters | 14.2 | 1.4:1 | Yes | 118″ |
| 17 Meters | 18.11 | 1.0:1 | Yes | 78½″ |
| 15 Meters | 21.15 | 1.1:1 | No | 99″ |
| 12 Meters | 24.94 | 1.2:1 | No | 78″ |
| 10 Meters | 28.8 | 1.2:1 | No | 64″ |

Figure 5 — One side of the antenna with retracted whip attached.

choke, you can use a 1:1 balun such as the one in Figure 6 that I made with a T200-2 toroid, a couple of feet of enameled wire and some odds and ends I had in my junk box.

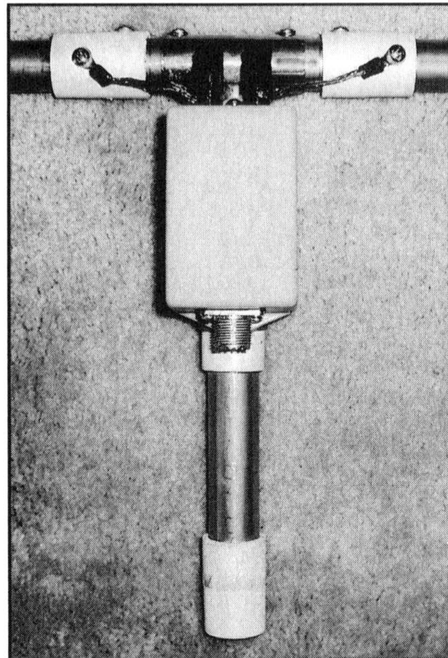

Figure 6 — Dipole feed connection — toroid balun option shown.

## Tuning it Up

Table 3 provides the data required to adjust the antenna for all five bands. You will likely want to adjust the length to whip at each noted frequency and record the length that provides a 1:1 SWR, since your construction might be different than mine. With the antenna adjusted to the frequency shown it should provide an acceptable SWR across each band, no greater than the *Max SWR* shown in the table.

If you were going to operate most of the time close to 14,000 kHz on CW and the band edge SWR of 1:4 is excessive for your radio, it can be corrected very easily by slightly lengthening the whip length to obtain a 1:1 SWR in the CW portion. On the other hand if you want to adjust for a 1:1 SWR at around 14,350 you can make the same type of a slight adjustment, this time by shortening the MFJ whips. By using this same technique on any band there is no need to bring along any bulky heavy antenna tuner on your camping adventure.

## Hoisting it Up

There are many ways of installing such an antenna in a portable environment. There are tripod mounts, improvised railing supports and many types of telescoping support poles available from a number of manufacturers. I have had good results using a Model S216 telescoping fiberglass pole manufactured by Hastings. These are available at electrical supply houses. There are other similar types sold by a number of Amateur Radio dealers.

In my usual installation, I have secured a short piece of CPVC tubing on the top telescoping section. On the other end of this short piece, I insert it into a CPVC coupling that is part of the T section on my antenna. I have also used short pieces of aluminum tubing fastened together made from old antenna beam elements. Even short pieces of inexpensive CPVC pipe will do for a 4 to 5 foot mast.

For the base, I often use a folding portable flood light base I picked up at a yard sale. If packing in by foot, I leave the base at home and use available supports such as tree stumps to hold up the mast by securing with a piece of small rope. With this method, one can set up this antenna within 5 minutes.

One final bit of advice — please be careful when setting up your antennas wherever you are. Before erecting any type antenna, always make it a habit to look for any electrical wires above and 360° around you.

*Clarke Cooper, K8BP, has been a ham since 1961 and holds an Amateur Extra class license. Clarke has a degree in electrical engineering and has spent most of his career designing manufacturing plants and automated control systems. He retired in 2001 as Project Manager for Federal Mogul.*

*He is a Life Member and a four term president of the Muskegon Area Amateur Radio Council in Muskegon, Michigan. He is also a member of the Thunderbird Amateur Radio Club in Phoenix, Arizona. He especially enjoys mentoring new hams and operating QRP on both CW and SSB.*

*Over the years, Clarke has had the pleasure of sending out over 30,000 QSL cards. He is a member of ARRL, QCWA, ARES, RACES, Fists and Adventure Radio Society. During the summer you can reach Clarke at 4202 E Pontaluna Rd, Fruitport, MI and in the winter at 17237 North 15th Pl, Phoenix, AZ 85022 or by e-mail at* **k8bp@earthlink.net**. QST

# A Portable Inverted V Antenna

*A portable antenna that is great for ARRL Field Day, offers directional control and may be useful where fixed antennas are not allowed.*

**By Joseph R. Littlepage, WE5Y**

If necessity is the mother of invention then restrictive covenants must be the mother of desperate measures. I recently moved into a new neighborhood and found that the covenants made it difficult for me to erect or install any type of ham radio antenna structure. If you are faced with a similar situation or simply have need for a portable antenna that you can erect quickly and easily, this may be what you need.

The problem I faced is common to many living in newer subdivisions and is quite restrictive of Amateur Radio operations. I was determined, however, to find a way to get back on the air. My challenge was to develop a portable antenna system that would be easy and quick to erect and enable me to work a reasonable amount of DX without being obvious to the neighbors. In addition, I needed to determine a low enough power level to avoid TVI and other potential interference that could draw attention to my Amateur Radio activity. QRP and a good antenna proved to be the answer!

## The Inverted V

I tried end-fed long wires, difficult-to-support dipoles and finally decided to try an inverted V cut for 17 meters. But how could I support it without a sky hook? That's when I hit upon the idea of using a lightweight telescopic pushup pole for a support mast. To serve as the base support for the pushup mast, I purchased a portable antenna tripod. I spread the tripod legs to about 40 inches apart and locked them in place. This combination is my sky hook!

The top of the antenna should bring together the feed line and two wire elements angled at least 90° apart. Each of the wire elements is cut for a $^1/_4$ $\lambda$ on the desired band. I chose to try the 17 meter band and cut each element for 18.1 MHz using the formula 234/f, with f the frequency in MHz for each quarter wave as shown in Table 1 for compromise CW and phone operation. You may wish to change the design frequency to suit your operational preferences. Final measurement and trimming was accomplished after the finished antenna was erected on site using an MFJ-259B antenna analyzer.

The next step was to devise a spreader assembly to hold the lower ends of the wire elements apart and keep the spread angle near 90° without resorting to ground anchor points. I wanted to be able to rotate the antenna using the "Armstrong" method to maximize its efficiency in selected directions. The solution was to use two 10 foot lightweight telescoping fiberglass fishing poles held end-to-end. A simple support arm assembly made of two 12 inch lengths of $^1/_2$ inch thin-wall PVC pipe joined to a $^3/_4$ inch PVC X connector. This holds the spreaders in a horizontal position and enables them to ride up and down on the main support mast. The free ends of the wire elements are attached to the eyelets on the outer ends of the two spreader poles that are extended to full length

**Table 1**

**Wire Half-Element Lengths, Portable Inverted V Antenna**

| *Band (Meters)* | *Design Frequency (MHz)* | *Length* |
|---|---|---|
| 20 | 14.175 | 16' 6 $^1/_8$" |
| 17 | 18.1 | 12' 11 $^1/_8$" |
| 15 | 21.175 | 11' $^5/_8$" |
| 12 | 24.94 | 9' 4 $^5/_8$" |
| 10 | 28.4 | 8' 2 $^7/_8$" |

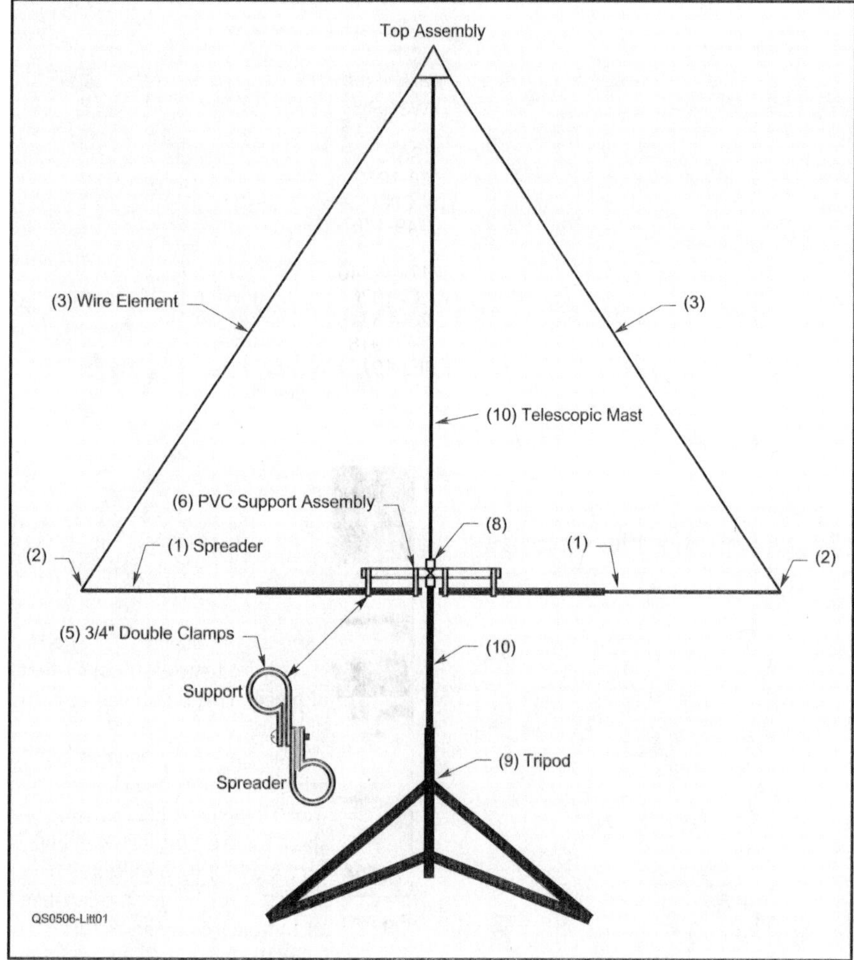


**Figure 1—General arrangement of the completed inverted V antenna.**

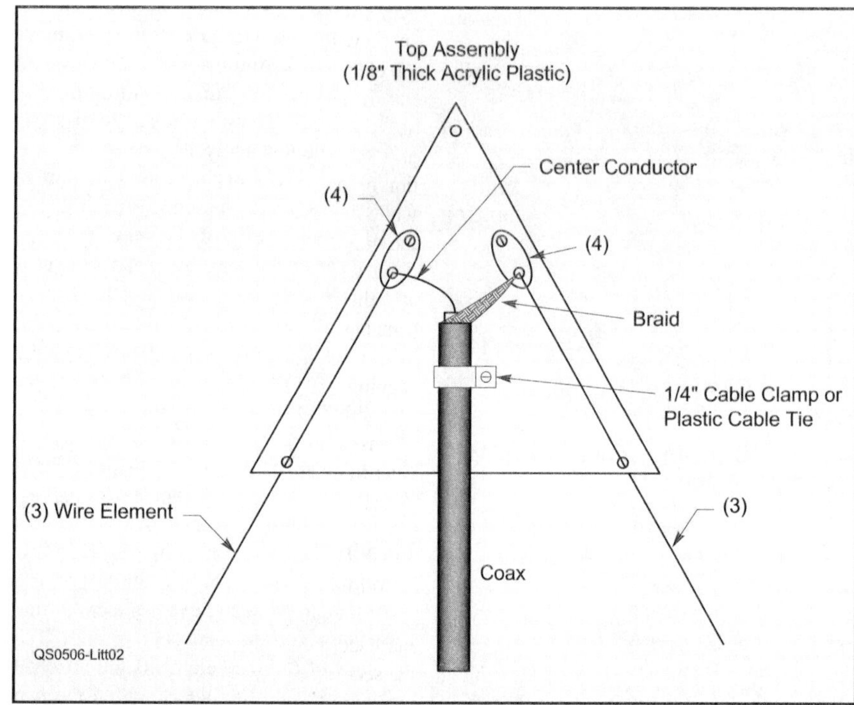


**Figure 2—Details of the top assembly.**

as shown in Figure 1. Table 2 provides the bill of materials and sources for all the antenna's components.

## Top Piece Assembly

A top fitting is required to bring the elements and transmission line together and to fix them to the top of the support. I constructed this assembly using a triangular piece of $^1/_8$ inch thick clear acrylic plastic or similar dielectric material. Its dimensions are not critical. Drill holes to accommodate a top hanger, two solder lugs, a nylon cable clamp for holding the coax and a hole in each lower corner to support the antenna wire. Neither the construction method nor the dimensions are critical.

Solder one end of an element wire and the center conductor of the coax to one solder lug and secure that to the top piece. Route the free end of that element half down through the lower hole on its side of the top piece. Solder the other element wire and the coax braid to the other solder lug and secure to the top piece in a like manner. Route that element half's free end through the other lower hole. Secure the coax to the top piece using a small nylon cable clamp as a strain relief. A small cable tie can be used instead if you carefully drill a hole on either side of the coax to route it through. The details are shown in Figures 2 and 3.

## Spreader Support Assembly

A $^3/_4$ inch PVC X connector with $^3/_4$ to $^1/_2$ inch bushings pressed into opposite side holes is used to hold the two 12 inch long support arms as shown in Figures 4 and 5. The support arms are cut from thin wall $^1/_2$ inch PVC pipe, used in preference to regular schedule 40 pipe to reduce weight.

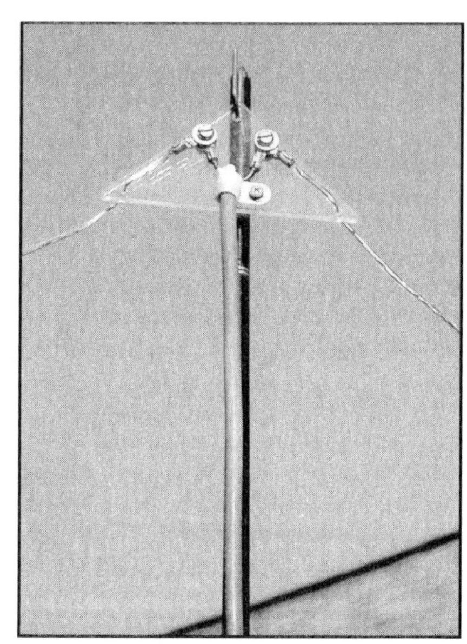

**Figure 3—Photo of the top assembly.**

**Table 2**
**Materials List for Portable Inverted V Antenna**

| *Item* | *Quantity* | *Description* | *Source* | *Part Number* | *Unit Price* |
|---|---|---|---|---|---|
| 1 | 2 | B'n'M Black Widow crappie fishing pole, 3 sections, 10 feet. | Wal-Mart | BW3RR | $ 8.50 |
| 2 | Package | Black interlock snap swivels, size 5. | Wal-Mart | BISS-5 | $ 0.50 |
| 3 | Roll | Insulated stranded #20 wire, 75 feet. | RadioShack | 278-1219 | $ 2.99 |
| 4 | Pack | Crimp-on ring tongues, 24 assorted. | RadioShack | 64-3030 | $ 1.49 |
| 5 | Pack | One-hole plastic clamp, 3/4" PP-1575UVB, six in pack. | Home Depot | E449-176 | $ 1.49 |
| 6 | 10 feet | Silver line (thin wall) PVC pipe, 1/2" | Home Depot | 717141340512 | $ 1.80 |
| 7 | 2 | PVC bushings, 3/4" × 1/2" | Home Depot | 1287162647 | $ 0.25 |
| 8 | 1 | PVC "X" fitting, 3/4" | Home Depot | 1287162483 | $ 1.14 |
| 9 | 1 | Portable antenna tripod. | MFJ | MFJ-1918 | $39.95 |
| 10 | 1 | Telescoping antenna mast, 33 feet. | MFJ | MFJ-1910 | $79.95 |

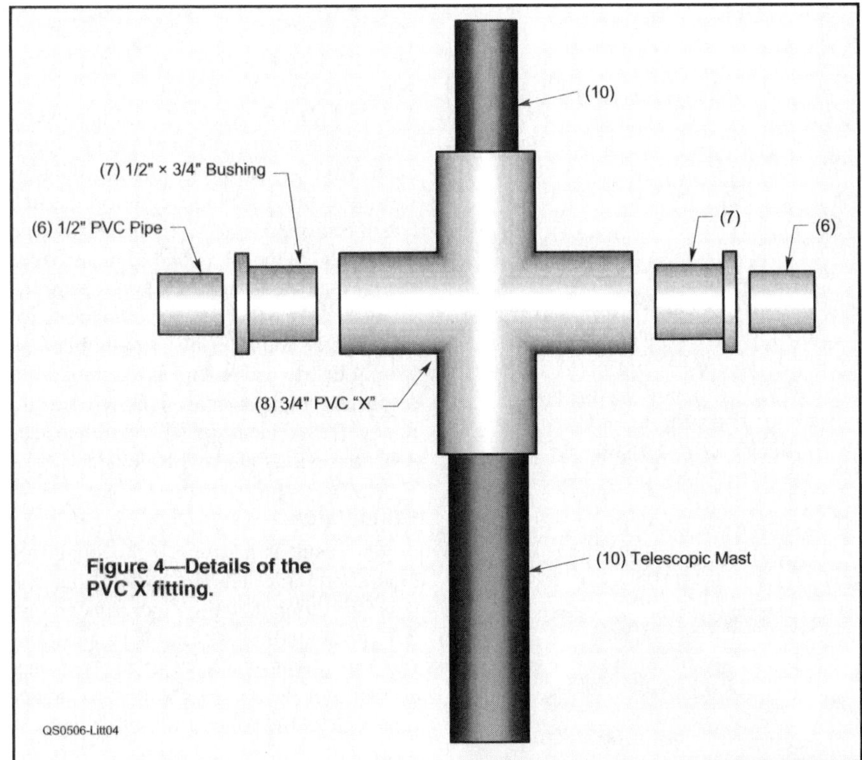


**Figure 4—Details of the PVC X fitting.**

**Figure 5—Photo of the PVC X fitting.**

Press the bushings into opposite holes of the X piece until they are firmly seated. Then firmly press a 12 inch support arm into each bushing. Cement all joints using PVC cement. The two remaining holes will allow the spreader assembly to ride up and down the vertical mast. A slight raised lip molded inside these holes should be dressed down flush with the inside surface using a round file. To each support arm install two sets of back-to-back cable clamps. One end of each set is firmly attached around the support arm and the other end of each set around a fishing pole spreader. The outer clamp sets should be placed very near the end of each 12 inch PVC support arm, as detailed in Figure 6.

## Stringing the Elements

When the top piece and the spreader support assemblies are completed it is time to string the elements. First, remove the innermost small top mast section with the eyelet from the support mast and set it aside. It will not be used. Place the large bottom end of the vertical mast, with the remaining nested sections inside, into the base supporting tripod. Next, raise the remaining centermost section of the telescopic support mast out about a foot and hold it in place with a spring-type clothespin.

Place the spreader support assembly down over the raised mast section with the spreader poles on the bottom side and against the clothespin. Next, attach the top assembly to the top opening of the small raised mast section with a bent hook fashioned from coat hanger wire run down inside the mast. The ends of the spreader poles can now be fully extended and twist locked in position. The free end of each wire element is then attached to the tip end eyelets on the spreader poles. Small fishing snap swivels can serve as attach hooks for the element ends.

Finally, remove the clothespin as you slowly start to raise the top mast section until it can be twist-locked into the next section below, which will follow it up. All the while the spreader support assembly will be moving down the vertical support mast to a point where the wire elements are straight and the spreader tips bend upward about 6 inches. This will maintain sufficient tension on the elements to prevent them from sagging. At this point, the spreader support assembly X connector should form a snug fit against the vertical mast. Failure to achieve a snug fit can result in the spreader assembly "weather vaneing" in a strong breeze.

## Erecting the Antenna

With everything set up it is simply a matter of raising the remaining vertical support mast sections and twist-locking each one to the section below it. You should take every precaution to prevent the mast sections from loosening and falling back down. This can

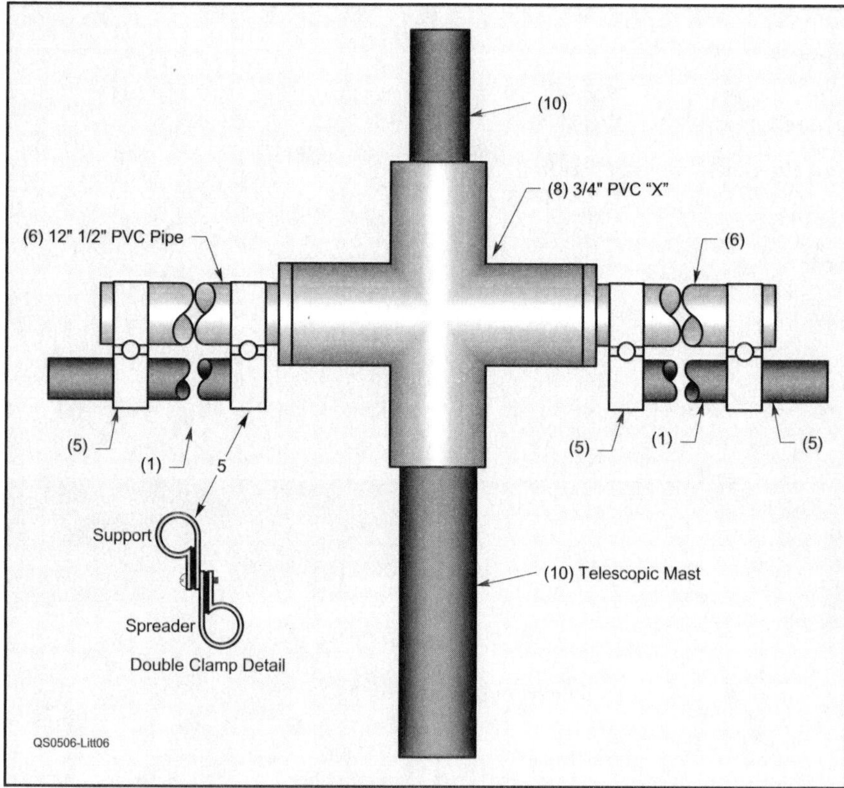


**Figure 6—Details of the spreader attachment method.**

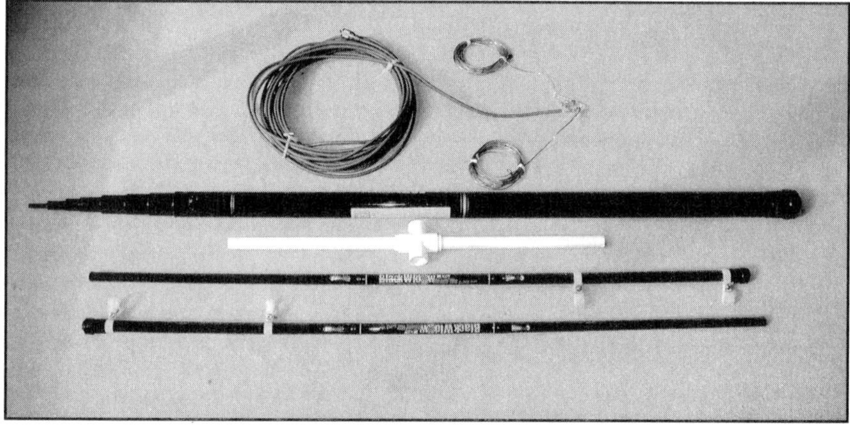

**Figure 7—The completed antenna broken down for storage or travel.**

result in fracturing the lower ends of the sections as they hit the bottom of the support tripod tube. Although the fiberglass sections are strong, they are thin and very brittle and will chip or crack when overstressed. Simply placing a couple of inches of cushioning material inside the bottom end of the tripod tube will prevent damage. The coax cable can be attached to the mast as it goes up. The final height depends on how many sections your support mast has. The pushup pole that I use has 11 sections. I use all but the smallest top section (with the small eyelet tip) since it is too weak to support the antenna assembly without bending over.

## Results

My first contacts were on 17 meter SSB running 5 W with an ICOM IC-703 powered by a 33 Ah gel-cell battery. A Connecticut station gave me a Q5 report during our 25 minute QSO. Two days later I worked Costa Rica, Guatemala, Colorado, New Mexico and Wisconsin with the same setup. Two of the QSOs were with pedestrian mobile stations.

The real test came a few weeks later when I worked ON6WA in Belgium and GI3DZE in Ireland using the same setup and power. I have received many good signal reports from the stations I have worked. When not in use, the antenna system can be broken down for storage or transport as shown in Figure 7.

For the 20 meter phone band, you can simply substitute a 1 inch PVC X connector with 1 inch to $^1/_2$ inch bushings and assemble the support arms as described above. This modification allows the entire spreader assembly to ride farther down the vertical support mast and thereby maintain the spreader tension on the longer 20 meter wire elements. As an alternative, you can add a 3 foot 6 inch long drop-down extension wire to the lower end of each 17 meter element and retune the antenna accordingly. Placing a small fishing weight on the free end of each extension wire will help to keep it vertical. For the 10, 12 and 15 meter bands the wire elements will be shorter than the 17 meter wire that I cut. To compensate for this, simply add a length of monofilament fishing line to each element to enable it to reach the spreader ends. This will ensure that the 90° element spread is maintained as well as tension on the wire elements.

## Conclusion

The result is a strong, lightweight, rotatable portable antenna system that can be easily constructed of inexpensive and readily available materials. In addition, you can elevate the apex of the V to any convenient height to clear an obstruction such as the edge of a roof or lower it to avoid neighborhood scrutiny. This antenna project has gotten me back on the air and could be the answer to some of your problems. The prospects for ARRL Field Day, emergency operations or backpacking to some remote site are limited only by your imagination.

*Joe Littlepage, WE5Y, was first licensed in 1969 as WB5AAI. He upgraded to Advanced Class in 1975 and more recently to Amateur Extra in 2002. He holds a degree in physics and is retired from the USAF. Joe is an active member of the Mississippi Coast Amateur Radio Association (MCARA) and serves as editor of the club newsletter. He enjoys QRP, kit building, homebrewing, building wire antennas and solar power experimentation. Joe can be reached at* **jlextra@netzero.net**.

By Markus Hansen, VE7CA

# A Portable 2-Element Triband Yagi

## Have you ever dreamed about a portable beam you could use at your summer cottage, while camping or on Field Day? Dream no longer. This portable beam can be rolled up and stashed in your car's ski boot!

Several years ago I entered the ARRL November Sweepstakes CW contest in the QRP category, operating from a portable location. It turned out to be a very frustrating experience with only 3 W of output power and dipole antennas. After the contest I decided that the next time I entered a QRP contest it had to be with gain antennas.

My philosophy has always been to try to keep life as simple as possible. In other words, I look for the easiest way to accomplish a goal that guarantees success. Don't get me wrong: Dipoles work particularly well considering the time and effort put into making them. But adding a reflector to a dipole antenna increases the overall gain about 5 dB, depending on the spacing between the elements. This extra gain makes a significant difference, especially when you are dealing with QRP power levels. My 3-W transmitted signal would sound like a 9.5-W powerhouse just by adding another piece of wire! And it would be inexpensive too.

With Solar Cycle 23 in full swing, having an antenna with gain on 15 and 10 meters also became a consideration. Another parameter was the sale of the family van, which meant the new antenna had to fit into the ski boot of our car. Keeping these constraints in mind, I used a computer antenna-modeling program, trying different design parameters to develop a triband 2-element

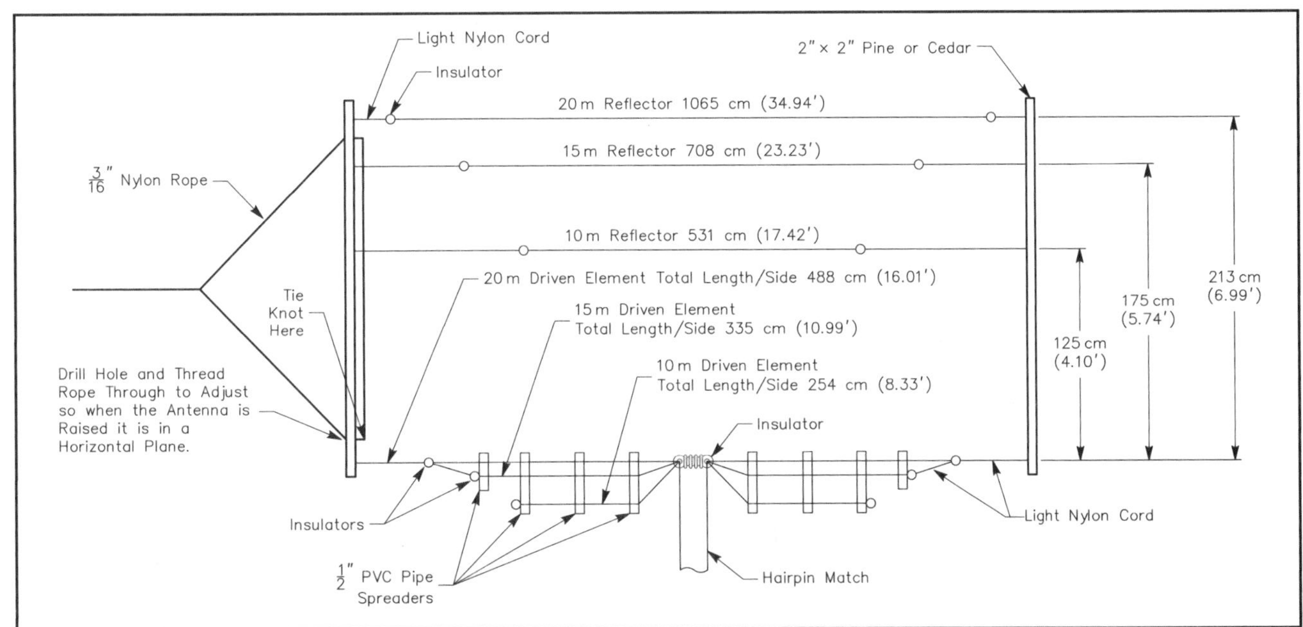


**Figure 1—Dimensions for VE7CA's 2-element wire triband Yagi.**

portable Yagi using wire elements.

The basic concept comprises three individual dipole driven elements, one each for 10, 15 and for 20 meters tied to a common feed point, plus three separate reflector elements. The elements are strung between two 2.13-meter (7-foot) long, 2×2-inch wood spreaders, each just long enough to fit into the ski boot of the car. Use the lightest wood possible, such as cedar, pine or spruce to keep the total weight of the antenna as light as possible. Fiberglass poles would also work, or PVC pipe reinforced with maple doweling to ensure they don't bend. (Wood has the benefit of being easy-to-find and very affordable).

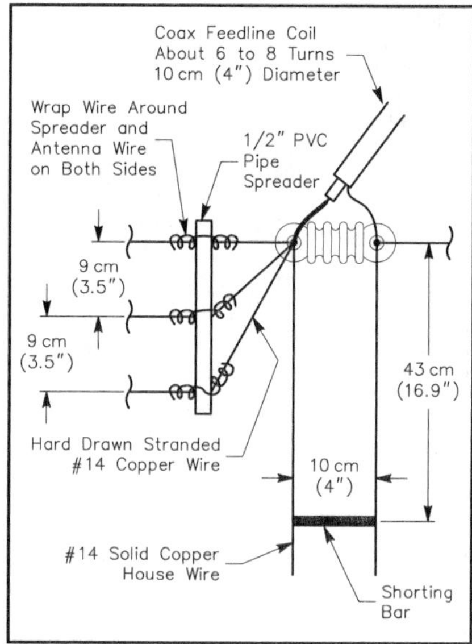


**Figure 2—Close-up view of the feed point.**

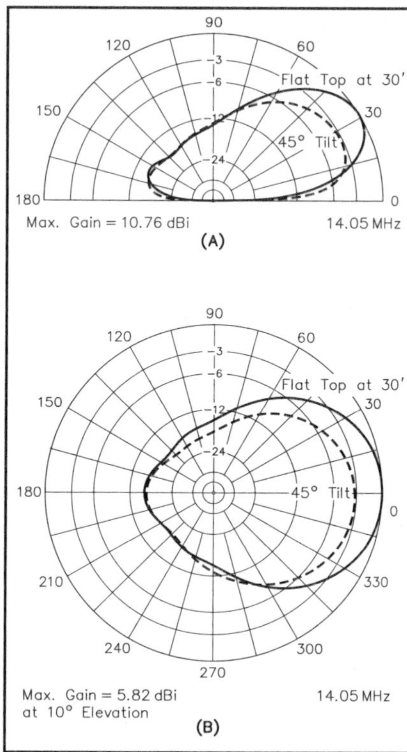


**Figure 3—At A, comparison of elevation patterns for VE7CA Yagi as a horizontal flat top (solid line) and tilted 45° from vertical (dashed line). At B, comparisons of azimuth patterns for a 10° elevation angle.**

Adding a reflector element relatively close to the driven elements lowers the feed-point impedance of the driven element, so a simple hairpin match was employed to match the driven elements to a 50-Ω feed line. Figure 1 shows the layout and dimensions of the antenna.

## The Hairpin Match

The matching system is very simple and foolproof. You should be able to copy the dimensions shown in Figure 2 and not need to retune the hairpin match, unless you plan to use the antenna in the top portions of the phone bands. The dimensions in Figure 2 produced a very low SWR—under 1.3:1 over the CW portions of all three bands. However, even in the lower portions of the SSB bands, the SWR doesn't rise above 2:1. SWR measurements were made at the end of a 25-meter (82-foot) length of RG-58 coax feed line.

Some may wonder why I used such a long feed line. First, when operating from a portable location it is better to be long than short. Nothing is more frustrating than finding that the coax you took along with you is too short. Further, when I change beam direction I walk the antenna around the antenna support, thus requiring a longer length than if I went directly from the antenna to the operating position.

If you are concerned about line loss you can run RG-58 down to the ground and larger-diameter RG-8 or RG-213 to the operating position. You may also find that in your particular situation a shorter length of coax will do. An 18-meter (59-foot) long piece of RG-58 has a loss of about 1 dB at 14 MHz, which is entirely acceptable considering the convenience of using coax cable.

### *Adjusting the Hairpin Match*

If after raising the antenna the SWR is not as low as you want in the portion of the bands you plan to operate, first double-check to make sure that all the elements are cut to the correct length and that the spacings between the driven elements and reflectors are correct. Next you can adjust the hairpin match. Connect either an antenna SWR analyzer or a transmitter and SWR meter to the end of the feed line and pull the antenna up to operating height. Determine where the lowest SWR is on 15 meters. By moving the shorting bar on the hairpin match up or down you can adjust the lowest SWR point to the middle of the portion of the 15 meter band you prefer. If your preference is near the top end of 15 meters you may have to shorten the 15-meter driven element slightly. After adjusting the 15-meter element and hairpin match, adjust the 10 and 20 driven-elements lengths separately, without changing the position of the shorting bar on the hairpin match.

The hairpin match is very rugged. You can attach the feed line to it with tape, roll it up, pack the antenna away and even with the matching wires bent out of shape it just seems to want to work.

### *Antenna Support*

Adhering to my constraint to keep things as simple as possible, I only use one support for the antenna, typically a tree. When the antenna is raised to its operating position it is a sloping triband Yagi. To achieve this, attach a rope to each end of the 2×2's to form a V-shaped sling, as shown in the Figure 1. Attach a length of rope to one sling and pull the antenna up a tree branch, tower or whatever vertical support is available. Tie a second length of rope to the bottom sling and anchor the antenna to a stake in the ground. By putting in two or three stakes in the ground around the antenna support, you can walk the antenna around to favor a particular direction. To change direction 180°, give the feed line a pull and the array will flip over. So simple but very effective!

### *Local or DX*

One of the features of a sloping antenna is that you can adjust the take-off (elevation) angle. For example, if you are interested in North American contacts (whether for casual QSOs or the ARRL SS contest), then sloping the antenna away sideways from the support structure at 45° with the feed point approximately 8 meters (26 feet) above the ground, will yield a 20-meter pattern similar to Figure 3A. Here, the maximum lobe is between 10° and 60° in elevation. The pattern of the antenna in a flat-top horizontal configuration at 9.1 meters (30 feet) is overlaid for comparison. You can see that the tilted beam has better low-angle performance, but at higher angles has less gain than its horizontal counterpart. Figure 3B shows an overlay of the azimuth patterns for these two configurations at a 10° takeoff angle.

If DX is your main interest, then you want to position the antenna even closer to vertical to emphasize the lower elevation angles. Figure 4 shows the pattern on 20 meters when the antenna is tilted sideward 10° away from vertical, again compared with the other orientations in Figure 3A. The feed point is 6 meters above ground and the model assumes fresh water in the far field, which is the case at my portable location.

Remember that the radiation pattern is quite dependent on ground conductivity and dielectric constant for a vertically polarized antenna. A location close to saltwater will yield the highest gain and the lowest radiation angle. With very poor soil in the near and far field, the peak radiation angle will be higher and the gain less.

I have had the opportunity to test this out at my portable location. Using two trees as supports, I am able to pull the antenna close to horizontal with the feed point about 7 meters above the ground. In this position, with 20 meters open to Europe, I have found it difficult to work DX on CW with 3 W of output power. However, when I change the slope of the antenna so that it is nearly vertical I not only hear more DX stations, but I find it relatively easy to work DX.

I have tried this many times, since it is simple to lower one end of the antenna to change the slope and hence the radiation take-off angle. The sloping antenna always performs much better for working DX than a low horizontal antenna. Recently, I worked nine European countries during two evenings of casual operating, even though the highest end of the antenna was only about 10 meters high, limiting the slope to about 45°.

Figure 5 shows the elevation pattern on 28.05 MHz for the beam sloped 10° from vertical at 45° from vertical, with the feed point at 8 meters height, again compared with the beam as a flat top at 9.1 meters (30 feet). With a steeper vertical slope, the 10-meter elevation pattern has broken into two lobes, with the higher-angle lobe stronger than the desired low-angle lobe.

This demonstrates that it is possible to be too high above ground for a vertically polarized antenna. Lowering the antenna so that the bottom wires are about 2.5 meters (8 feet) above ground (for safety reasons) restores the 10-meter elevation pattern without unduly compromising the 20-meter pattern.

### *Portable It Is*

A winning feature of this antenna is that it is so simple to put up, take down, transport and store away until it is needed again. When I am finished using the antenna and it's time to move on, I just lower the array and roll the wire elements onto the 2×2's. I put a plastic bag over each end of the rolled-up array and tie the bag with string so that the wires don't come off the ends of the 2×2's. I then put it in the ski boot of a car, or in the back of a family van and away we go. At home, it takes very little space to store and it is always ready to go—No bother, no fuss.

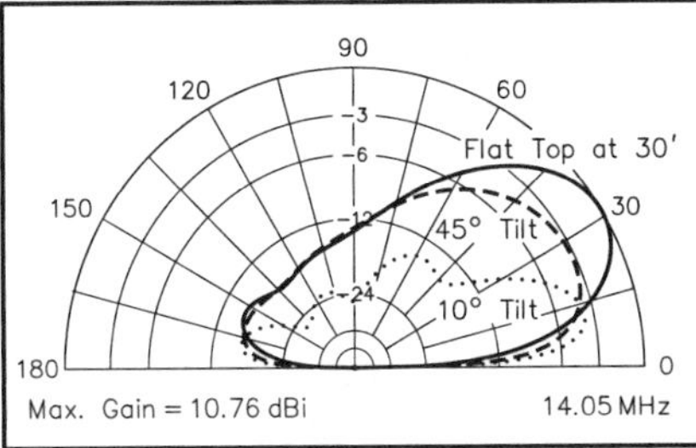


**Figure 4—Comparison of elevation patterns for VE7CA Yagi as a horizontal flat top (solid line), tilted 45° from vertical (dashed line) and tilted 10° from vertical (dotted line).**

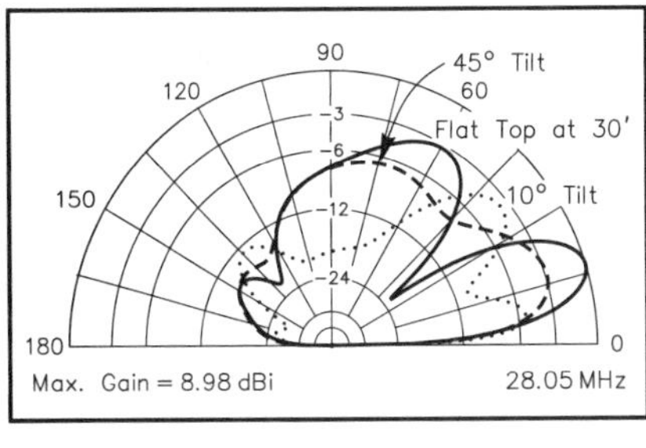


**Figure 5—Same antenna configurations as shown in Figure 4, but at 28.05 MHz. On 10 meters, the flattop configuration is arguably best, but the 45° tilted configuration is not far behind.**

### *Testimonial*

How well does it work? It works very well. On location I use a bow and arrow to shoot a line over a tall tree and then pull one end of the array up as far as possible. For DX I aim for a height of 20 to 30 meters if possible. For the Canada Day, Field Day and Sweepstakes contests I aim for a height of about 15 meters. This antenna helped me to achieve First Place for Canada, in the 1997 ARRL CW Sweepstakes Contest, QRP category.

The ability to quickly change direction 180° is a real bonus. Late in the 1997 ARRL SS CW contest with the antenna pointed east I tuned across KH6ND. He was the first Pacific station I had heard during the contest and obviously I needed to work him. After trying many times to break through the pileup and not succeeding, I flipped the antenna over to change the direction 180° and then worked him on my next call. Figure 6 shows the azimuth pattern at 21.05 MHz for the beam mounted with a 10° slope from vertical. There is a very slight skewing of the azimuthal pattern because the slope away from purely vertical makes the antenna geometry asymmetrical.

VE7NSR, the North Shore Amateur Radio Club, has used this antenna sloped at about 45° for the last two years on 20 and 10 meters on Field Day with good success. The title photo shows the antenna attached to a tower during Field Day.

As they say, the proof is in the pudding. If you need a 20 to 10 meter antenna with gain, this has to be one of the simplest antennas to build, and it will work every time!

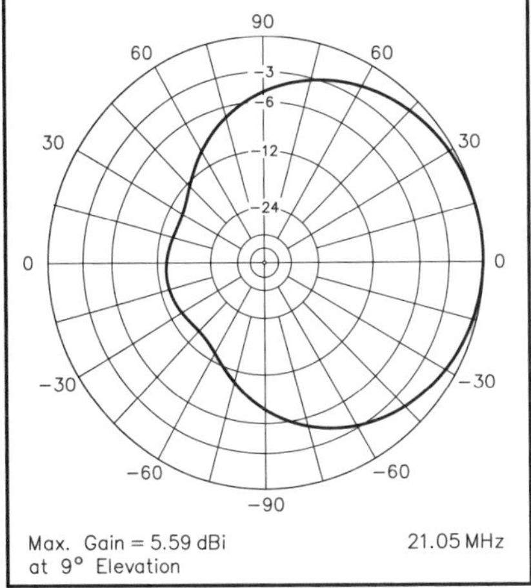


**Figure 6—Azimuthal pattern for VE7CA Yagi tilted 10° from vertical on 15 meters.**

*Markus Hansen, VE7CA, was first licensed as VE7BGE in 1959. He has been a member of ARRL since he received his license. His main interests include DX, collecting grids on 6 meters, contesting and building his own antennas and various types of ham-radio equipment. He is also an ardent CW operator. Markus has had two previous articles published: "The Improved Telerana, with Bonus 30/40 Meter Coverage," in* The ARRL Antenna Compendium Vol 4 *and "Two Portable 6-Meter Antennas" in* The ARRL Antenna Compendium Vol 5. *You can contact Markus at 674 St Ives Cres, North Vancouver, BC V7N 2X3, Canada, or by e-mail at* **ve7ca@rac.ca**.

*You can download the EZNEC input-data files as* ***VE7CA-1.ZIP*** *from ARRLWeb* **(www.arrl.org/files/qst-binaries/)**.

By Phil Salas, AD5X

# A Simple HF-Portable Antenna

Tired of dragging that bulky old antenna tuner along on your vacation jaunts? Spare your suitcase and your pocketbook because this simple multiband wire antenna will get you on the air in a jiffy—with no extra gear required.

Every summer my wife (N5UPT), my daughter (AC5NF) and I spend about a week on Mustang Island off the coast of Corpus Christi, Texas. I always enjoy operating HF-portable when on vacation, and because Mustang Island is also known as IOTA NA092 (Islands On The Air, North American island number 92), getting on the air is even more fun! In case you're imagining typical DXpedition fare, you should know right from the start that we don't exactly rough it on Mustang Island. In fact, we always stay in a condo, which I request to be "the highest one available."

My first portable rig was a Kenwood TS-50, followed by an MFJ-9420 (see May 1999 *QST*). Last year I went deluxe and upgraded to an ICOM IC-706MKII. That little rig works dc to light—all bands and all modes, with goodies to boot. It it is an excellent choice for almost any type of portable operation.

I've experimented with several types of antennas on these outings—including Hamstick mobile whips, resonant dipoles and random-length wire dipoles fed through a tuner. I prefer resonant antennas so I don't have to worry about transporting and storing an antenna tuner. Of course, multiple dipoles or a handful of Hamsticks can take up a lot of room.

Last summer I used the multiband dipole described here with excellent results. If you're interested in a simple multiband wire that's easy to build and pack away, give this antenna a try.

The basic antenna covers all bands from 20-10 meters. You could increase its coverage, but the dimensions of a typical condo balcony seem to limit the lower frequency to 20 meters or so. If your operating site is larger, feel free to scale the antenna appropriately.

Basically, the antenna started as a full-size 20-meter dipole. I then inserted small in-line insulators to allow for multiband operation as shown in Figure 1.

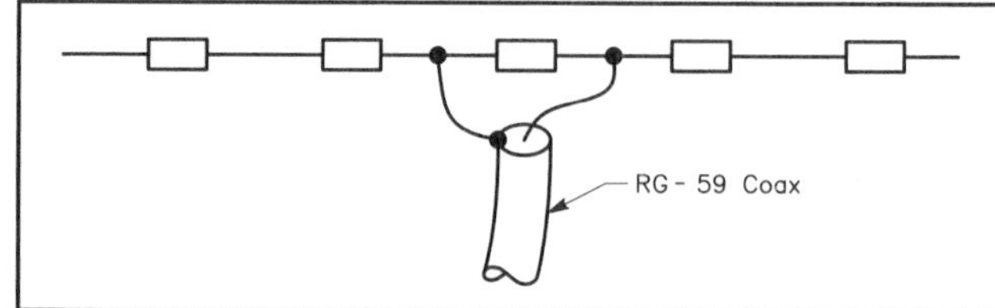


**Figure 1—The concept began with a full-size dipole antenna that I "broke up" with small insulators.**

The insulators are $^3/_8$-inch (diameter) by one-inch nylon spacers that can be found at most hardware stores. Each spacer is used as a "band switch" by drilling a small hole in each end and threading a short length of #14 bare wire (house wire) through each end, and attaching a short piece of wire terminated in an alligator clip. The clip, shown in Figure 2, is available at RadioShack stores (ask for part number 270-380).

I used #24 insulated wire for the dipole elements because it's lightweight and flexible. Obviously, any type of wire is fine. Use whatever you have on hand. The best way to determine the various segment lengths is to calculate the individual dipole lengths using:

L (feet) = 468/freq (MHz)

Tack solder the wire sections to the insulators, attach a feed line (RG-59 coax will do) and hang the dipole in a convenient place where it's easy to work on and adjust. Although the SWR meter method will work, to adjust the multiband dipole properly, beg, borrow or buy an antenna analyzer.

First, "unclip" all of the alligator clips and adjust the inner wire segments for the lowest SWR on your favorite part of the 10-meter

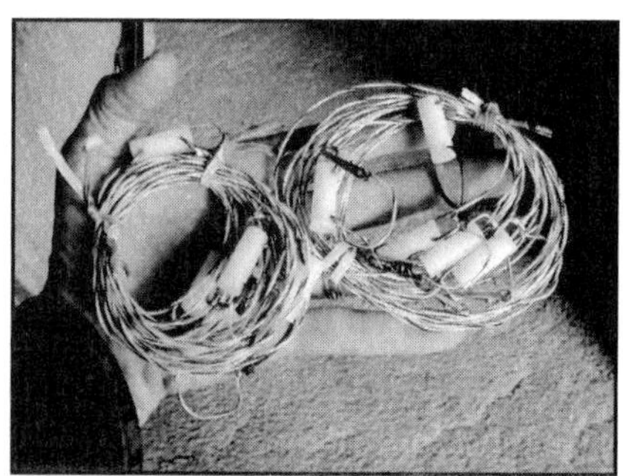

**The entire antenna can be collapsed to a size that fits in the palm of your hand!**

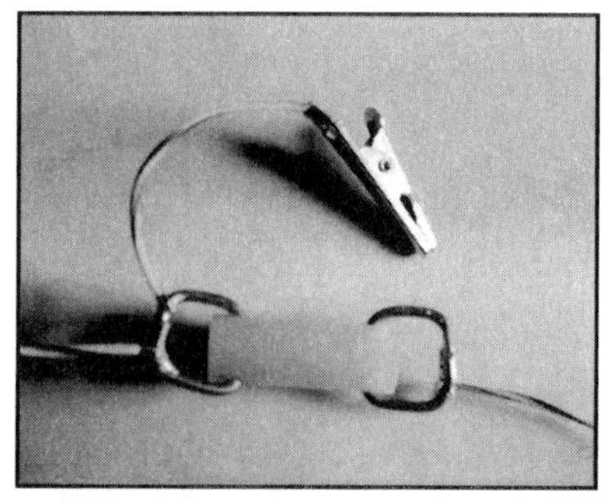

**Figure 2—The band switches are constructed from nylon spacers, some wire and an alligator clip.**

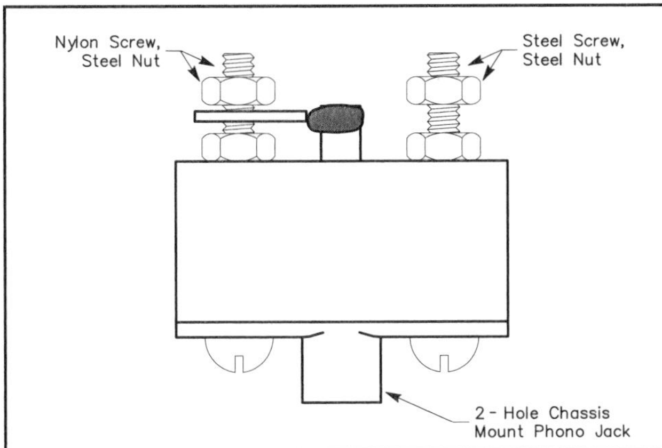


**Figure 3 — I used an extra nylon spacer for the center insulator. I drilled the ends and attached a chassis-mount phono jack as shown. The nylon screw is used on one side to make sure that the phono jack's center conductor doesn't short to ground. I soldered #4 spade lugs to the inside ends of the 10-meter dipole elements so the dipole can be easily attached (and detached) to the center insulator. Feel free to use other center insulator designs as desired.**

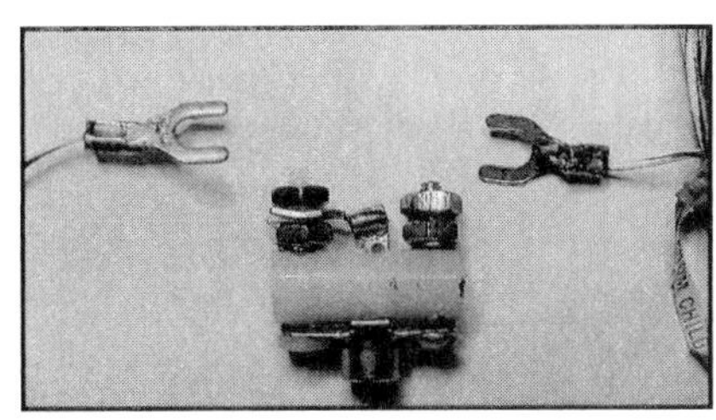

**A photograph of my version of the center insulator.**

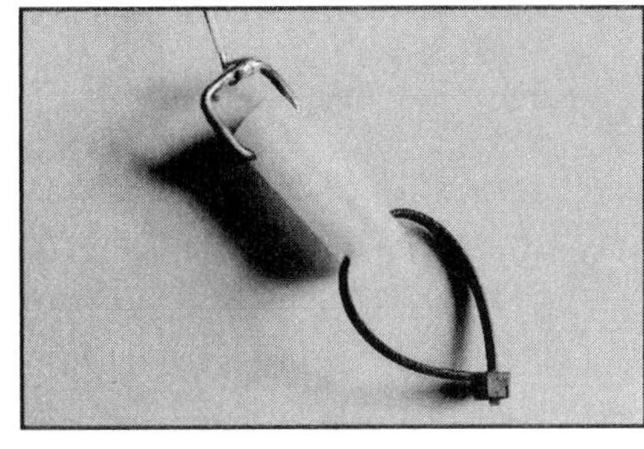

**My design for the end insulator.**

band. The wires should be a bit long, so unsolder them on one end and trim them as follows:

New length = Original length × Measured low-SWR Frequency/ Desired low-SWR frequency

Next, clip (attach) the inner pair of alligator clips and adjust the next segment for resonance on 12 meters using the formula and steps described previously. Continue this procedure for 15, 17 and 20 meters.

I know—you're adjusting your antenna low to the ground and your particular portable mounting location will undoubtedly vary. For our purposes it really doesn't matter. Most modern rigs can put out full power into a 2:1 SWR, so reasonable location-based SWR variations probably won't affect your rig's operation. If the SWR is really high, something's drastically wrong or you have the alligator clips set up for operation on the wrong band, etc. Incidentally, you can use a balun if you want to. I normally don't worry about feed line transformers when operating portable.

The antenna leg lengths I wound up with are shown below:

10 meters: 8 feet 3 inches on each side

12-10 meters: 10 inches on each side

15-12 meters: 1 foot 4 inches on each side

17-15 meters: 1 foot 8 inches on each side

20-17 meters: 3 feet 9 inches on each side

Each side is a total of 15 feet, 10 inches, for a total of 31 feet, 8 inches for the entire antenna.

Finally, if you want to electrically "shorten" your antenna, make the clip lead wires a little longer and wrap the excess wire around the insulators to make loading coils.

I used an extra nylon spacer for the center insulator. I drilled the ends and attached a chassis-mount phono jack as shown in Figure 3. The nylon screw is used on one side to make sure that the phono jack's center conductor doesn't short to ground. I soldered #4 spade lugs to the inside ends of the 10-meter dipole elements so the dipole can be easily attached (and detached) to the center insulator. Feel free to use other center insulator designs as desired.

## Conclusion

If you need a simple portable antenna, spend an hour or two assembling this one. It's simple, cheap and a good performer. Simply adjust the clip leads for the desired frequency band and you're on the air—no tuner required! Sure, you have to make a quick trip to the balcony (or whatever) to change bands…but this is a vacation-oriented design, after all!

**1517 Creekside Dr**
**Richardson, TX 75081**
**ad5x@arrl.net**

# The Shooter — A 3-Band Portable Antenna

Want a vertical you can put up or take down in a jiffy? Try this penny-pincher's Field Day special!

By E. W. "Twisty" Ljongquist,* W4DWK/W1CQS

When Wayne Stump, WB4ARZ, brought his little marvel over for me to see, I was admittedly more than a bit skeptical. His previous accounts of the lightweight antenna had aroused my curiosity. Could an antenna made from the miscellany found around the house or in the neighborhood live up to what I'd been told?

The day came when Wayne arrived with his compact bundle of hardware. I looked at the collection and then him, gesturing as I laughingly remarked, "you gotta be kidding!" His only answer was a smile as he set about assembling the various pieces of metal that were transformed in a matter of minutes into a vertical antenna. "Too flimsy to be practical," I told myself as he put on the final touches.

When I moved closer to the antenna, however, I realized I'd judged the book by the cover — for I found the construction to be quite sturdy. "Well, what do you think of it now?" he remarked.

My reply, "Not too bad an idea for the

*1855 Meridian Rd., West Palm Beach, FL 33409

Fig. 1 — Wayne Stump, WB4ARZ, seated at this outdoor radio location in sunny Florida. His version of the vertical Shooter antenna is visible at the left.

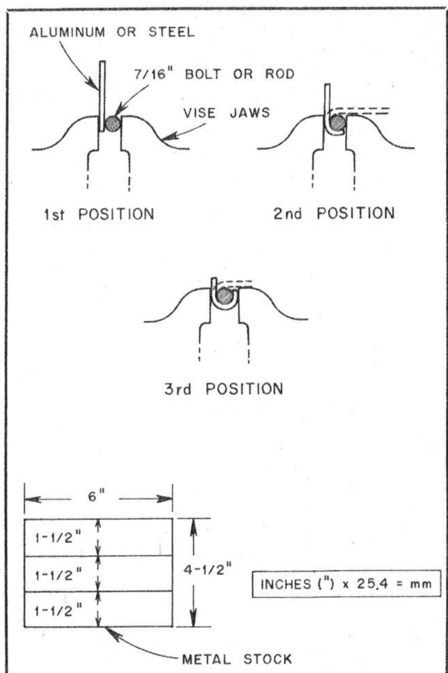


Fig. 2 — Dowels for the Shooter are made by placing the metal stock (shown at the bottom) in a vise along with a 7/16-inch (11-mm) dia bolt or rod. The metal is bent into a cylindrical shape as indicated by the three drawings. The dowels can be made from either aluminum or steel.

chap who enjoys jaunts to the 'boonies' or DXpeditions. As a matter of fact, I'd say it's nearly tailor made for someone like you who has a travel trailer!"

Wayne could sense that my doubts were rapidly disappearing. As I grasped the antenna and gave it a little tug, I declared with a tone of approval, "What a nifty vertical for Field Day. Even condominium of apartment dwellers can use it!"

### A Simple Evening Project

Once the material has been gathered, the work of making this antenna should take little more than an hour. You might say it makes a good project for a rainy evening.

Look at Fig. 1, which shows Wayne at the key of a portable installation. You will see one of these vertical antennas positioned just to the left of the operating bench. The scene is rather typical of a basic Field Day station. Figs. 2, 3 and 4 provide details of the Shooter.

In knocked-down form, the entire package is very compact. The maximum length of the antenna segments is such that there should be no problem carrying the bundle or even shipping it by airplane. I estimate the weight to be about six pounds. All parts, except the conduit, can be fitted into a 6-inch cube and still leave enough spare room to accommodate a small L network. In my opinion, a portable antenna like this is nearly ideal, especially if you consider the fact that the total cost should be no more than $7.

Moments after Wayne left, I began a scavenger hunt around my QTH. That adventure netted most of the components needed to construct a copy of his vertical. Naturally, the odds and ends I found were not all identical to the parts he'd acquired. But the resulting variations I made in building the Shooter in no way impaired the performance.

The main ingredients of the antenna are two 10-foot (3.05-m) lengths of 5/8-inch (15.9-mm) OD thin-wall electrical conduit.[1] These may be purchased at an electric supply store or a place that handles building materials. Two couplers, suitable for that diameter conduit, are also required along with a washer that has a 5/8-inch opening. At the time of this writing a 10-foot length of this conduit sells for about $2.

To cut the conduit into the prescribed lengths, I recommend the use of a pipe cutter. Tools of this type are available at many hardware stores. Although a hacksaw will serve the purpose, a pipe cutter will provide a more even cut.

### Preparing and Fastening the Sections

As you begin the construction of the antenna, cut one of the 10-foot lengths in half. Cut the other length at 6 feet (1.83 m), then provide a length of 27 inches (686 mm) and another of 6 inches (152 mm). File off the burrs and lay this material aside.

Three metal dowels are to be made. These aid in holding together the sections of the antenna, besides serving as a means for electrical continuity. I formed my dowels from a fairly stiff piece of aluminum chassis material that was 6 inches (152 mm) wide and long enough so that guidelines may be drawn on the metal as explained below.

Dowel dimensions are presented in Fig. 2. With the help of a square, draw the three guidelines across the piece of aluminum so that the lines are 1-1/2 inches (38 mm) apart.

Now obtain a bolt or rod, 7/16 inch (11 mm) in diameter, for shaping the dowel. The length of the bolt or rod should equal the length of the dowel. Clamp the aluminum stock and the bolt in a vise with the bolt just catching the edge of the aluminum. Fig. 2 shows how this is done. Bend the stock around the bolt. Release the vise and reposition the aluminum so that it can again be bent further around the bolt once the pair has been reclamped in the vise. After this step, you should have a U-shaped end on the aluminum stock. Now cut off the straight section of the aluminum with snips or a hacksaw. Reclamp the U-section and the bolt in the vise and finish bending the dowel into a cylindrical shape with the help of a hammer and block of wood. With a little judicious work using pliers, you should then have a dowel that fits properly in the conduit. Fashion the remaining dowels in the same manner.

To begin putting the antenna sections together, insert one of the dowels into a 5-foot (1.52-m) section and another in a 6-foot (1.83-m) section. What you do with the third dowel is explained in Table 1.

Secure each dowel with a self tapping screw. This screw should be placed 1-1/2 inches (38 mm) from the end of the conduit. Two thin-wall compression couplers are to be installed as shown in Fig. 4. These reinforce the joints and aid the dowels in keeping the antenna straight.

As you may have gathered from Fig. 1, this vertical antenna is not self-supporting. Guys are clearly visible in the picture. Wayne and I guy our respective antennas in a slightly different manner. Because he has welding equipment, he brazed three wire loops on one of the compression nuts. Each loop serves as a

[1]Notes appear on page 70.

Fig. 3 — This is all there is to the Shooter except the pegs. When the antenna is being stored or transported, a good idea is to keep the guys and radials coiled and tied with plastic ties.

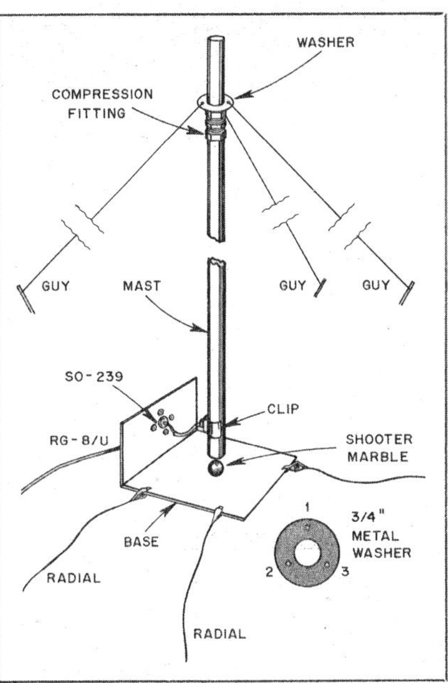


Fig. 4 — The Shooter is mounted and guyed as indicated by this illustration. A metal washer is drilled at points 1, 2 and 3 (equidistant from each other), and a guy is connected to each hole. The washer slips over a section of the conduit coming to rest atop one of the compression fittings. Ground radials are clipped to the metal base.

terminal for a guy. On the other hand, I simply drilled three holes in a 3/4-inch (19-mm) flat washer which slides over the conduit. You can see how this is done by referring to Fig. 4. A guy is attached to each of the three holes.

Wayne chose plastic clothesline, but I wanted something lighter for guys. To save weight and storage space, I preferred a strong grade of nylon fish line that has a thickness about equal to that of a mason's or carpenter's chalk line. Where the antenna is to be installed at ground level, tent pegs are practical for anchoring the bottom ends of the guy lines.

### The Shooter Base

Now, you may wonder why I dubbed my version of the vertical antenna the Shooter. Well, actually this "moniker" is related directly to what I use for the base insulator. Ever play "mibs" as a youngster? Remember the "glassies" or "shooters" that a flick of the thumb sent dashing toward your opponent's marbles? Well, that is exactly what serves as the base for my antenna. (I found one tucked away in a bureau drawer.) Of course you could use a cone-shaped stand-off insulator instead, but I liked the idea of the glassie. Mine is epoxied to the base to prevent it from getting lost.

Whereas Wayne fabricated a base plate from flat washers and a steel peg, one made in the following manner may be easier to fabricate. Aluminum stock that is left over from making the dowels could be cut to provide a 4 × 7 inch (100 × 175 mm) plate that can be bent at right angles 3 inches (75 mm) from one end. A 3/8-inch (9.5-mm) hole may be drilled in the center of the bottom of the base and a 3/4-inch (19-mm) hole placed in the upright part to accommodate the coaxial fitting. The base you see in Fig. 3 actually is a bracket from an old amplifier. Originally it held an amplifier tube in place at the socket.

Mount an SO-239 coaxial connector on the base. Solder a short length of flexible wire to the center conductor of the connector. The length should be such that you can make contact with the bottom of the antenna by means of a clip. Such a clip may be salvaged from an old fuse block that was a part of a main distribution box removed from house wiring. Solder the wire to the clip and then slide the clip onto the antenna. An alternative might be to bend a piece of spring-brass material to an equivalent shape.

### A Vertical Needs Radials

A quarter-wave vertical antenna will not function properly without radials. In general, the more radials you provide, the better the signal will be. Ground rods introduce considerable ground loss, and for that reason are not considered effective rf grounds. I use only four radials, mainly for the sake of portability. Because of my location over the dry Florida sands, however, I should increase the number.

**Table 1**

**Radiator Lengths**

| *Band* | *Length* |
|---|---|
| 10 meters | A 6-foot section plus the 27-inch piece. |
| 15 meters | A 6-foot section plus one 5-foot length. The third (loose) dowel can be used to change length if necessary. |
| 20 meters | A 6-foot section, two 5-foot lengths plus a 6-inch piece. Use the loose dowel for connecting the top section. |

While touching on the subject of radials, I do recommend a review of the excellent articles concerning ground systems and SWR written by both Jerry Sevick, W2FMI[2] and M. Walter Maxwell, W2DU.[3] A good resume of Maxwell's series is contained in the book, *Amateur Radio Techniques,* published by the RSGB.

I find much truth in the statement that SWR isn't everything. For instance, when I adjust my antenna, I frequently find that minimum SWR and maximum field strength do not coincide. In my opinion, a simple field-strength meter will tell you and me more about antenna performance than any SWR indicator.

A valid reason for having an SWR indicator, though, is that it will let you know if your rig will "like" what it sees as it looks into the transmission line. Some of the new transceivers have a nasty habit of popping out when the SWR goes over 2:1. Who needs that to happen? To compensate for mismatch and any reactance presented by the antenna system, an antenna tuner (more properly called an impedance-matching network) is desirable. The popular Transmatch circuit or even a simple L network should handle any matching requirements concerned with this portable vertical antenna. What the tuner does is to provide an adjusted load that is appropriate for a 50-ohm transmitter output. The tuner does not change a mismatch that may be present at the antenna feed point. For the portable vertical to have the best feed-point match in relation to the transmission line, a good radial system and adjustment of the antenna height are required.

### Radiator Lengths for Each Band

To select the correct radiator elements for each of the three bands, choose the lengths from Table 1. Those dimensions are taken from *73 Vertical, Beam and Triangle Antennas,* an excellent antenna book written by Edward Noll, W3FQJ.[4] Those dimensions are for the cw portions of the respective bands. They can be found in Chart 2 on page 11 of Noll's book. You may disregard the transmission-line lengths suggested by Noll in connection with this antenna. As George Grammer, W1DF, says, "A line is just to carry the rf to the antenna." Very seldom, and only in special cases, is the line anything but that.

Although the antenna dimensions I've listed are mainly for the cw portions of each of the three bands, I find that with these same measurements operation in the phone segments is quite satisfactory. Only at the highest end of the bands did the SWR get up to 3:1. The signal reports I received while operating my FT-101B with the Shooter assured me that the signals are getting out.

Putting this antenna up is easy once you get used to the procedure. Lay the proper assembly (the elements for the particular band you've chosen) on the ground with the lower end approximately at the point of the erection. Install the compression couplings and tighten them snugly with pliers. Fasten two of the guys, allowing some slack. Incidentally, my guys are 16 feet (4.9 m) long. With the third guy in your hand and the third peg within reach, raise the antenna for a visual check. If the antenna is nearly vertical, set the third peg and fasten the bottom of the third guy. Should the antenna be too far out of plumb, you'd do better to lower it and rearrange it.

Once the radiator is up, lift the antenna enough so that the base can be slipped beneath, allowing the antenna to come to rest atop the marble. Then shift the base as needed to make the antenna plumb.

With the antenna in proper position, connect the coaxial transmission line to the SO-239 connector on the base. Lay out the radials in a random manner, clipping each one to the base as shown in Fig. 4. You are now ready to fire up the rig and make the tuning adjustments.

Tuning consists of setting the antenna tuner for minimum SWR, adjusting the conduit length if necessary (I did not have to make a length change) and observing the field-strength meter for maximum signal output. If you find that maximum signal strength does not occur at the same point as minimum SWR, do not be overly concerned. In that case a compromise will provide you with the best results.

Both Wayne and I hope that those amateurs who have pleaded for information about simple portable antennas will try the Shooter. We feel the small amount of effort required to make one will be well spent. It is not a real DX chaser, as compared to a quad mounted 60 feet above ground, but you can have fun with it. QST

**Notes**

[1]Electricians call this "half-inch conduit."

[2]Sevick, "The W2FMI Ground-Mounted Short Vertical," *QST,* March 1973, and "Short Ground-Radial Systems for Short Verticals," *QST,* April 1978.

[3]Maxwell, "Another Look at Reflections," Parts 1 through 7, *QST,* April, June, August and October 1973, April and December 1974, and August 1976. [No additional parts for this series are planned for *QST* appearance. — Ed.]

[4]Published by Editors and Engineers (Howard Sams).

# A Traveling Ham's Trap Vertical

## If you're a wayfarer, you should be interested in this two-band trap vertical, which collapses to 39 inches for easy transport.

By Doug DeMaw,* W1FB

The traveling ham's trap vertical ready for final assembly. At the left foreground are two wing-shaped pieces of aluminum used experimentally during development of the antenna; they are not required for the model described in the text.

How effective is a trap type of antenna? That's a question the ARRL staff is asked repeatedly. Unfortunately, there is no simple answer we can offer. This is because the term "effective" is rather subjective. Effectiveness for one operator might mean the ability to work rare DX. Conversely, another amateur might consider the antenna effective if it helped him to break into pileups quickly. Some other operator might think of antenna effectiveness as a quality that would permit maintaining reliable communications over a specified ground-wave path.

Perhaps a better question to ask would be, "How *efficient* is a trap style of antenna?" But even that query is a tough one to address. The efficiency depends on many factors, such as the quality of the ground system, the Q of the traps (minimum losses) and the tuning of the system. The antenna performance also will depend on its height above nearby conductive or absorptive objects.

The best response to questions about trap-antenna performance is probably something like, "In theory, a full-size antenna will perform better than a short antenna with lumped constants." In other words, there is a certain trade-off to be expected in any "compromise" antenna. The performance degradation of a compromise antenna may, in some instances, be so minor that the operator would never recognize the difference between it and a full-size antenna. On the other hand, the performance difference can be startling. An example of the latter was seen during a DXpedition by W8JUY/8P6WM and W1FB/8P6EU. The two stations had equal transmitter power and were situated 100 feet (30.5 m) apart. The antennas of each station were erected over the ocean shore. W8JUY used a 4-band trap vertical at a height of 40 feet (12 m) above the sea shore. A complete radial system (factory specified) was erected for the vertical antenna. W1FB used a center-fed sloping dipole over the sea shore. It was approximately 25 feet (7.6 m) above the shoreline. During 20-meter operation over a two-week period, comparative signal reports from the USA and DX stations revealed that the sloping dipole of W1FB was consistently two to three S units better than the trap vertical. It is fair to say that at another time, and from some new QTH, the situation might be reversed. The inferior performance of the trap vertical did not, however, impair the ability of W8JUY to hold regular schedules or work DX, which brings us to the focal point of this discussion: Irrespective of the actual performance, a trap beam or trap vertical can be (and usually is) a suitable all-around amateur hf-band antenna.

**Table 1**

**Dimensions for Various Frequency Pairings**

| *Tubing (Fig. 1)* | *Band (MHz)* | *A* | *B* | *C* | *D* | *E* | *F* | *C1 (pF)*** | *L1 (Approx. μH)* |
|---|---|---|---|---|---|---|---|---|---|
| Tubing Length (inches) | 21/28 | 25 | 16 | 25 | 25 | 25 | 33 | 18 | 1.70 |
| | 14/21 | 38 | 33 | 37 | 37 | 37 | 33 | 25 | 2.25 |
| | 10/14* | 42 | 42 | 54 | 54 | 54 | 49 | 39 | 3.25 |
| Tubing Length at Resonance (approx. inches)*** | 21/28 | 20 | 16 | 21.5 | 21.5 | 21.5 | 33 | — | — |
| | 14/21 | 33 | 33 | 33 | 33 | 33 | 33 | — | — |
| | 10/14* | 37 | 37 | 49.6 | 49.6 | 49.6 | 49 | — | — |

in. × 25.4 = mm
*New WARC-79 band.
**See text.
***Midband dimensions. $X_{C1}, X_{L1} \approx 300\ \Omega$

*Senior Technical Editor, ARRL

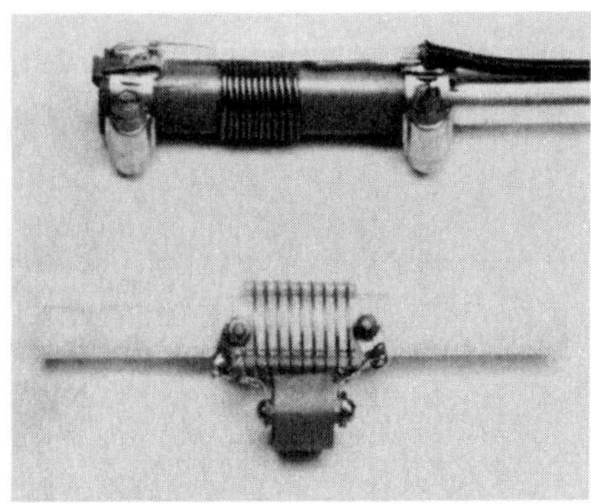

Fig. 1 — Photograph of two styles of traps. PVC tubing is used for the lower one (see text) and Teflon rod is used as the coil form for the other. The ends of the Teflon rod have been reduced on a lathe to fit inside the tubing sections above and below the trap. Teflon is too flexible for large verticals, but phenolic or fiberglass rod would be satisfactory. Miniductor stock comprises the coil in the upper trap. A ceramic transmitting capacitor is used in parallel with the coil.

Separate single-band, full-size vertical antennas would probably work slightly better than a multiband trap vertical, but the latter would cost less, occupy less space, be more convenient and do a pretty good job for the operator.

### How do Traps Work?

A *trap* is pretty much what the name implies: It "traps" rf energy by blocking its passage to portions of the antenna that aren't used during operation in a specified band. A trap is a high-impedance device at the selected operating frequency, which enables it to block the passage of rf energy on the band for which it is a parallel-resonant circuit. At some lower frequency it becomes part of the antenna, and responds somewhat as a loading coil. Thus, if we were to build a two-band trap beam or vertical for, say, 40 and 20 meters, the part of the antenna from the feed point to the bottom of the trap would constitute the 20-meter radiator. The trap (tuned for 20 meters) would effectively "divorce" the portion of the antenna above the trap during 20-meter operation. When changing operations to 40 meters, all of the antenna would become a functional part of the system. The trap acts as a loading coil on 40 meters, which results in the overall antenna length being somewhat less than that of a full electrical quarter wavelength.

If the trap antenna were designed for more than two amateur bands, additional traps and antenna sections would be used, but the principle of operation would remain the same. The highest frequency section of the antenna is always nearest the feed point and progresses outward until all of the antenna is used for the lowest operating frequency.

### Trap Design

There is no rigid formula for selecting a best L/C ratio for an antenna trap. Generally, the $X_L$ and $X_C$ values can range between 100 and 300 ohms, and since $X_L = X_C$ at resonance, they will be the same value in a trap.[1] We often select a standard capacitor value in that reactance range and "tailor" the coil $X_L$ to equal it.

It is important to keep the trap losses as low as possible (high Q). This requires the use of a coil form with high dielectric quality, such as polystyrene, phenolic, fiberglass, ceramic, Teflon or Plexiglas. The material chosen should be capable of withstanding antenna stress during periods of wind and ice loading. Brittle materials should not be used in cold climates in order to prevent the coil form from shattering under stress.

For power levels below 200 watts dc input to the PA stage of the transmitter, it is okay to use PVC tubing. At high rf-power levels, the PVC may heat, melt and burn. Nylon insulating forms are subject to the same failure, and the condition worsens as the operating frequency is increased.

The coil conductor should be of large cross-sectional area to minimize trap loses. No. 14 AWG or heavier copper wire is recommended. For high-power work, it is helpful to use copper tubing as the coil.

The Q and voltage rating of the trap capacitor also is an important consideration. If a fixed-value capacitor is used, it should be a ceramic transmitting type of capacitor (Centralab 800 series or equivalent). These units are available with working voltages up to 20 kV. This style of capacitor is shown in Fig. 1.

A suitable length of 50- or 75-ohm coaxial cable can be used as a trap capacitor. RG-58/U and RG-59/U cable is suggested for rf power levels below 150 watts. RG-8/U or RG-11/U will handle a few hundred watts without arcing or overheating. The advantage in using coaxial line as the trap capacitor is that the trap can be adjusted to resonance by selecting a length of cable that is too long, then trimming it until the trap is resonant. This is possible because each type of coax exhibits a specific amount of capacitance between the conductors. *The Radio Amateur's Handbook* contains a table that lists the capacitance per foot for popular coaxial cables. For example, RG-58/U has approximately 25 pF per foot, whereas RG-8/U cable has approximately 30 pF per foot. A coax-cable capacitor is also shown in Fig. 1.

### Additional Band Capability

Assuming that we have built a trap vertical for two or three bands, what might we do to obtain capability for one additional band without using the trap concept? The simple way to achieve this is to place a coil and capacitance hat on the top of the trap vertical. This is equivalent to "top loading" any vertical antenna. Suppose we had a trap vertical for 20, 15 and 10 meters, but also wanted to use the system on 40 meters. The popular way to do this is to construct a 40-meter loading coil which can be installed as shown in Fig. 2. A number of commercial trap verticals use this technique. The loading coil is called a "resonator" because it makes the complete antenna resonant at the lowest chosen operating frequency (40 meters in our example). The coil turns must be adjusted while the antenna is assembled and installed in its final location. The remainder of the antenna has to be adjusted for proper operation on all of the bands *before* the resonator is trimmed for 40-meter resonance. If the capacitance-hat wires are short (approximately 12 inches or 300 mm), we can assume a capacitance of roughly 10 pF, which gives us an $X_C$ of 2275 ohms. Therefore, the resonator will also have an $X_L$ of 2275 ohms. This becomes 51 $\mu$H for operation at 7.1 MHz, since $L_{\mu H} = X_L/2\pi f$. The resonator coil should be wound for roughly 10% more inductance than needed to allow some leeway for trimming it to resonance. Alternatively, the resonator can be wound for 51 $\mu$H and the capacitance-hat wires shortened or lengthened until resonance in the selected part of the 40-meter band is achieved.

As was true of the traps, the resonator should be wound on a low-loss form. The

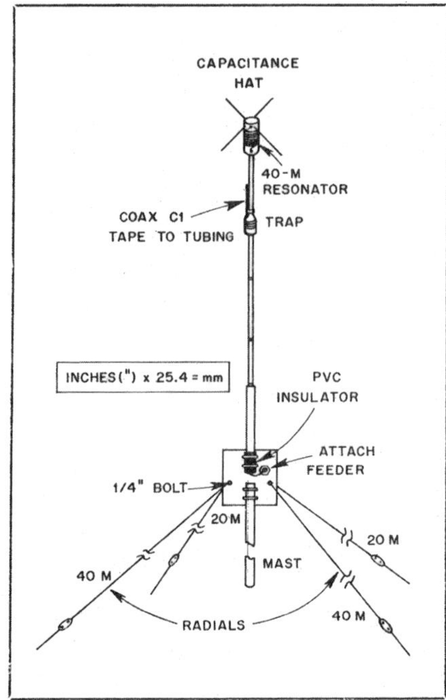


Fig. 2 — The assembled trap vertical, showing how a resonator can be placed at the top of the radiator to provide operation on an additional band (see text).

[1]Editor's Note: This is true only for one frequency band. All other dimensions being held equal, the lower the L/C ratio, the higher the resonant frequency will be on a lower frequency band. Of course the length of the outside end section can be adjusted for desired resonance on the lowest frequency band with a fixed L/C ratio.]

largest conductor size practical should be used to minimize losses and elevate the power-handling capability of the coil. Details of how a homemade resonator might be built are provided in Fig. 3. The drawing in Fig. 2 shows how the antenna would look with the resonator in place.

### A Practical Two-Band Trap Vertical

The author needed a 20/15-meter antenna for use on his RV camper and to carry to the Caribbean for DXpeditions. Therefore, it seemed prudent to design an antenna that could be broken down into a small package for carrying or storage. The best approach seemed to be that of using short lengths of aluminum tubing that would telescope into one another. The longest of these is 39 inches (991 mm).

The ends of the sections are cut with a hacksaw to permit securing the joints by means of stainless-steel hose clamps. The trap is held in place by two hose clamps that compress the PVC coil form and the 1/2-inch (13-mm) tubing sections onto 1/2-inch dowel-rod plugs (Fig. 4). Strips of flashing copper (parts "G" of Fig. 4) slide inside sections B and C of the vertical. The opposite ends are placed under the hose clamps, which compress the PVC coil form. This provides an electrical contact between the trap coil and the tubing sections. The ends of the coil winding are soldered to the copper strips. Silicone grease should be put on the ends of straps "G" where they enter tubing sections B and C. This will retard corrosion. Grease can be applied to all mating surfaces of the telescoping sections for the same reason.

The trap (after final adjustment) should be protected against weather conditions. A plastic drinking glass can be inverted and mounted above the trap, or several coats of high-dielectric glue (Polystyrene Q-Dope) can be applied to the coil winding. If a coaxial-cable trap capacitor is used, it should be sealed at each end by applying noncorrosive RTV compound.

The trap is tuned to resonance prior to installing it in the antenna. It should be resonant in the center of the desired operating range, i.e., at 21,050 kHz if you prefer to operate from 21,000 to 21,100 kHz. Tuning can be done while using an accurately calibrated dip meter. If the dial isn't accurate, locate the dipper signal using a calibrated receiver *while the dipper is coupled to the coil and is set for the dip.*

A word of caution is in order here: Once the trap is installed in the antenna, it will not yield a dip at the same frequency as before. This is because it becomes absorbed in the overall antenna system and will appear to have shifted much lower in frequency. For the 20/15-meter vertical, the apparent resonance will drop some 5 MHz. Ignore this condition and proceed with the installation.

### The Tubing Sections

An illustration of the assembled two-band trap vertical is shown in Fig. 5. The tubing diameters indicated are suitable for 15- and 20-meter use. The longer the overall antenna, the larger should be the tubing diameter to ensure adequate strength.

A short length of test-lead wire is used at the base of the antenna to join it to the coaxial connector on the mounting plate. A banana plug is attached to the end of the wire to permit connection to a uhf style of bulkhead connector. This method aids in easy breakdown of the antenna. A piece of PVC tubing slips over the bottom of section "F" to serve as an insulator between the antenna and the mounting plate.

If portable operation isn't planned, fewer tubing sections will be required. Only two sections need be used below the trap, and two sections will be sufficient above the trap. Two telescoping sections are necessary in each half of the antenna to permit resonating the system during final adjustment.

### Other Bands

It's unlikely that everyone would want to build this antenna for 20 and 15 meters. Those who are interested in other frequency pairings will find pertinent data in Table 1. We have included information on building a trap vertical for the new 10-MHz WARC-79-sanctioned band, plus 20 meters. One additional band can be accommodated for any of the combinations shown by using a top resonator.

### A Universal Mounting Plate

Field operation requires a variety of antenna-mounting schemes such as porch-rail attachment, window-sill mounting,

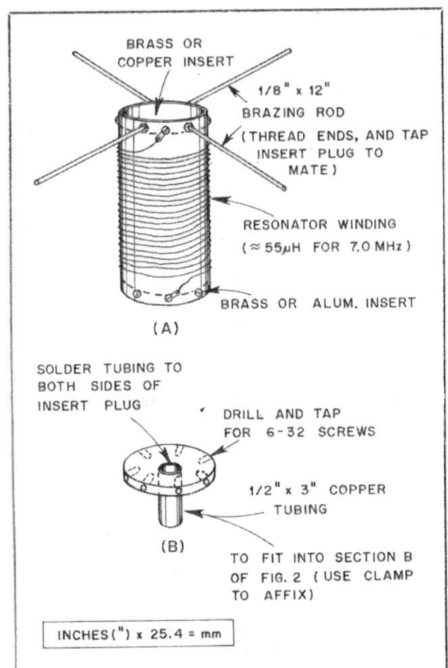


Fig. 3 — Details for building a homemade top-loading coil and capacitance hat. The completed resonator should be protected against the weather to prevent detuning and deterioration.

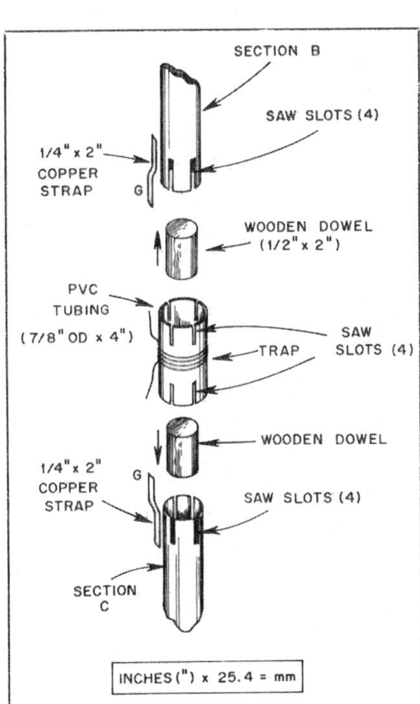


Fig. 4 — Break-down view of the PVC trap. The hose clamps that go on the ends of the PVC coil form are not shown.

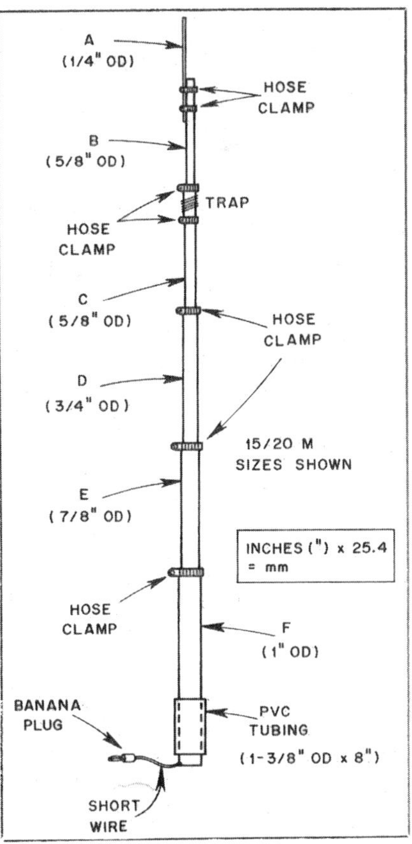


Fig. 5 — Assembly details for the two-band trap vertical. The coaxial-cable trap capacitor is taped to the lower end of section "B."

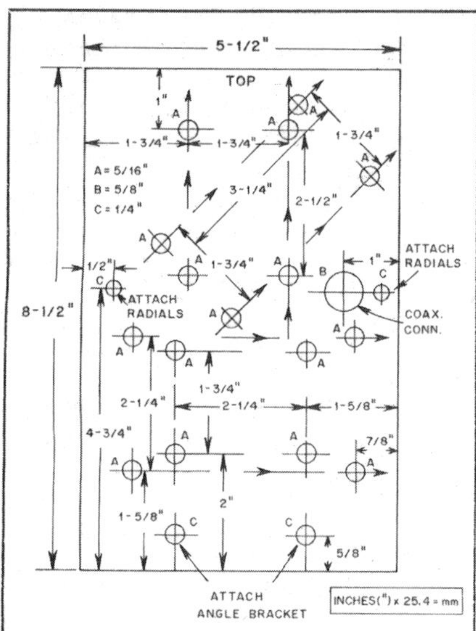


Fig. 6 — Suggested layout for a universal mounting plate. U bolts or muffler clamps can be used to attach the vertical to the plate, and to affix the supporting mast to the antenna. The arrows show the angles at which the antenna and the mast can be mounted to meet a host of mechanical conditions. Hole "B" is for a uhf bulkhead coax connector.

etc. For this reason it is helpful to have a mounting plate which will satisfy many unknown conditions. Fig. 6 shows the layout for a plate that is made from 1/4 inch (6.3 mm) thick aluminum plate. Steel, copper or brass plate is also suitable. The top half of the plate contains two sets of U-bolt holes for mounting the radiator vertically. A second set of holes permits 45-degree mounting from a window sill.

The lower half of the plate has two sets of U-bolt holes for attaching a mast vertically. A second set of holes permits horizontal mast mounting, should that format be necessary.

The bottom two holes in the plate are for attaching a length of angle stock. There is a second piece of angle stock the same length. The angle brackets can be used with a pair of C clamps to attach the antenna to a variety of foundations.

Hole "B" is for the uhf female-to-female coaxial bulkhead connector. Holes "C" at the left and right center of the plate are for attaching the radials by means of large bolts and washers. The holes marked "A" are for the U bolts or muffler clamps. The hole diameters and spacings will depend upon the size and brand of U bolts used. The arrows indicate the mounting angle of the antenna element and the mast. A photograph of the base plate and related hardware is given in Fig. 7.

### Ground System

There's nothing as rewarding as a *big* ground system. That is, the more radials the better, up to the point of diminishing returns. Some manufacturers of multiband trap verticals specify two radial wires for each band of operation. Admittedly, an impedance match can be had that way, and performance will be reasonably good. So during temporary operations where space for radial wires is at a premium, use two wires for each band. We prefer a minimum of four wires for each band, and generally use that many. The slope of the wires will affect the feed-point impedance. The greater the downward slope, the higher the impedance. This can be used to advantage when adjusting for the lowest VSWR. When the radials are perfectly horizontal, the feed impedance will be on the order of 30 ohms. This suggests the use of a 1.6:1 broadband transformer at the feed point to assure a good match — assuming horizontal radials must be used in a particular installation.

The radial wires are cut to an electrical quarter wavelength for the band of operation. Some operators argue that making the radials 5% longer will increase the antenna bandwidth. The author has not found this to be true, but did observe some changes in feed-point impedance when that was done.

If the antenna is to be ground-mounted, make certain that the lower end of section "F" is only a few inches above ground. Bury as many radials in the earth as practical, using no less than 20 wires that are the length of the overall vertical antenna, or longer.

### Tune-up

An SWR indicator will be necessary when adjusting the antenna. Apply rf

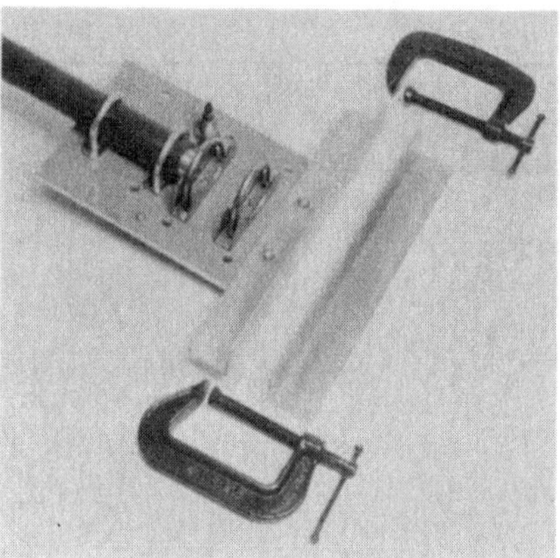

Fig. 7 — Photograph of the mounting plate showing two angle brackets and C clamps, which provide additional mounting possibilities.

energy at the center of the desired operating range of the highest operating band of the antenna. Adjust the length of the section below the trap for the lowest VSWR.

Next, set the transmitter for a frequency in the center of the lowest frequency operating range. Adjust the length of the section above the trap for the lowest obtainable VSWR. Repeat both adjustments to compensate for interaction of the adjustments. If a top resonator is used for a third band, it should be adjusted last for the lowest attainable VSWR.

### Summary Comments

The antenna can be broken down to form a compact assembly for transport. A heavy-duty cardboard mailing tube, or a 2-inch (51-mm) ID piece of aluminum tubing will serve nicely as a container for shipping or carrying. Iron-pipe thread protectors can be used as plugs for the ends of the carrying tube. The trap, mounting plate and coaxial feed line should fit easily into a suitcase with the operator's personal effects.

If you haven't been a radio-operating wayfarer thus far, perhaps this antenna will inspire you to become one! If you want to hear this antenna in operation, look for W1FB on 20 and 15 meters from the camp site, or from Tortola, British Virgin Islands during late October and early November of 1980.

By Allen Baker, KG4JJH

# The Black Widow—A Portable 15 Meter Beam

Looking for a portable 15-meter antenna for camping or Field Day? Four fishing poles, 50 feet of wire, a few pieces of wood, some PVC and a painter's pole make a stinging portable performer.

## The Lure

There are a lot of excellent antenna designs available to the amateur, but few can combine light weight and portability with the significant gain and front-to-back ratio (F/B) of the Moxon Rectangle. All the materials necessary for building this portable antenna (including the mast) are available from your local Wal-Mart, Home Depot and RadioShack for under $125.

## The Moxon Rectangle

The Moxon Rectangle is a two-element beam with considerable gain, high F/B ratio, and a direct match to 50 Ω coax. [The Moxon-type antenna is named for its originator, L. A. Moxon, G6XN.—*Ed.*] The antenna is a derivative of the Moxon Square and it has been further optimized by L. B. Cebik, W4RNL.[1] Figure 1 shows the basic dimensions of the antenna.

I used the program *MoxGen*[2] to generate a model at 21.070 MHz for use with PSK and then fine-tuned it with *EZNEC*.[3] Since the Black Widow uses insulated wire (and *MoxGen* is based on bare wire), the final dimensions will change due to the insulation's effect on the wire's velocity factor.

I first built the antenna to the wire lengths in the *EZNEC* model: A = 203.25″, B = 31.25″, C = 4.88″ and D = 37.63″, which resulted in an actual resonant frequency of 20.180 MHz. I trimmed the wires to a final length = FM/FD × L, where FM is the measured frequency, FD is the desired frequency and L is the length before trimming. The final dimensions are A = 194.63″, B = 29.88″, C = 4.75″ and D = 36.0″. For an antenna centered in the middle of the 15-meter band: A = 193.75″, B = 29.75″, C = 4.63″ and D = 35.88″.

## Construction

Figure 2 presents an overview of antenna assembly. All of the necessary materials are listed in Table 1 along with sources for each. The antenna can be fashioned from the explanations contained here; however, detailed construction drawings for the antenna and the mast components are available at **www.arrl.org/files/qst-binaries/black-widow.zip**.

[1]Notes appear on page 79.

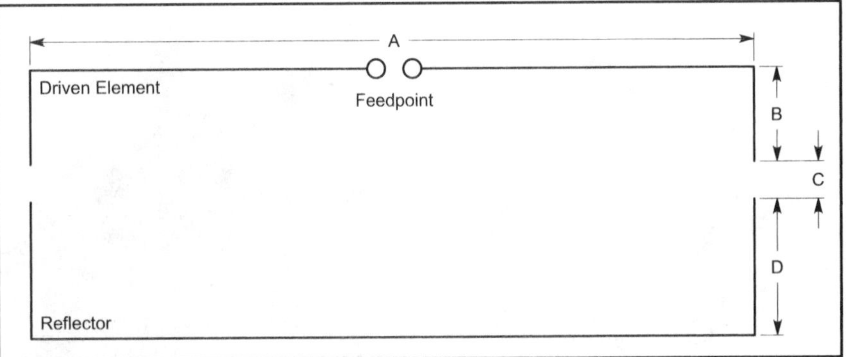


**Figure 1—The basic dimensions of the Moxon Rectangle.**

**Table 1**
**Materials List for the Black Widow Antenna**

| *Item* | *Qty* | *Description* | *Source* | *Price Each ($)* |
|---|---|---|---|---|
| 1 | 4 each | Crappie Master Fishing Pole, RCM-10LW, 3 sections, 10 ft | Wal-Mart | 6.96 |
| 2 | 1 roll | 20 AWG insulated stranded wire, 75 ft | RadioShack 278-1219 | 2.99 |
| 3 | 1 each | Terminal strip, 8-position | RadioShack 274-678 | 2.49 |
| 4 | 1 each | Connector, SO-239 | RadioShack 278-201 | 1.99 |
| 5 | 1 pkg | Soldering ring, #6 | RadioShack 64-3030A | 1.69 |
| 6 | 1 each | Epoxy, quick set | Home Depot | 2.97 |
| 7 | 4 each | Screw hook, 0.162″ wire diameter | Home Depot | 0.42 |
| 8 | 1 each | Wood, poplar, ½″ × 6″ × 36″ | Home Depot | 5.25 |
| 9 | 2 pkg | Three piece extension pole, Linzer RP503 | Home Depot | 3.79 |
| 10 | 1 each | Painter Pole, Mr. LongArm Model 2324 | Home Depot | 39.97 |
| 11 | 1 each | PVC coupling, 1½″ | Home Depot | 0.47 |
| 12 | 1 each | CPVC pipe, ½″ × 10 ft | Home Depot | 2.98 |
| 13 | 1 each | CPVC drop ear 90° elbow, ½″ | Home Depot | 0.68 |
| 14 | 1 each | CPVC 90° elbow, ½″ | Home Depot | 0.19 |
| 15 | 10 each | Sheet metal screw, self-drilling, #6 × ½″ | Home Depot | 0.17 |
| 16 | 230 ft | Nylon twine, #36 | Home Depot | 3.79 |
| 17 | 3 each | Clothesline tightener, Lehigh 7097 | Wal-Mart | 2.96 |
| 18 | 50 ft | Nylon rope, $^3/_{16}$″ | Wal-Mart | 1.97 |
| 19 | 1 each | 2″ × 4″ × 8′, pine | Home Depot | 3.00 |
| 20 | 2 each | L-bracket, 3″ × 3″ × ¾″ | Home Depot | 3.00 |
| 21 | 5.5 ft | RG-8X coaxial cable | RadioShack 278-1313 or equiv | 0.32 per ft |

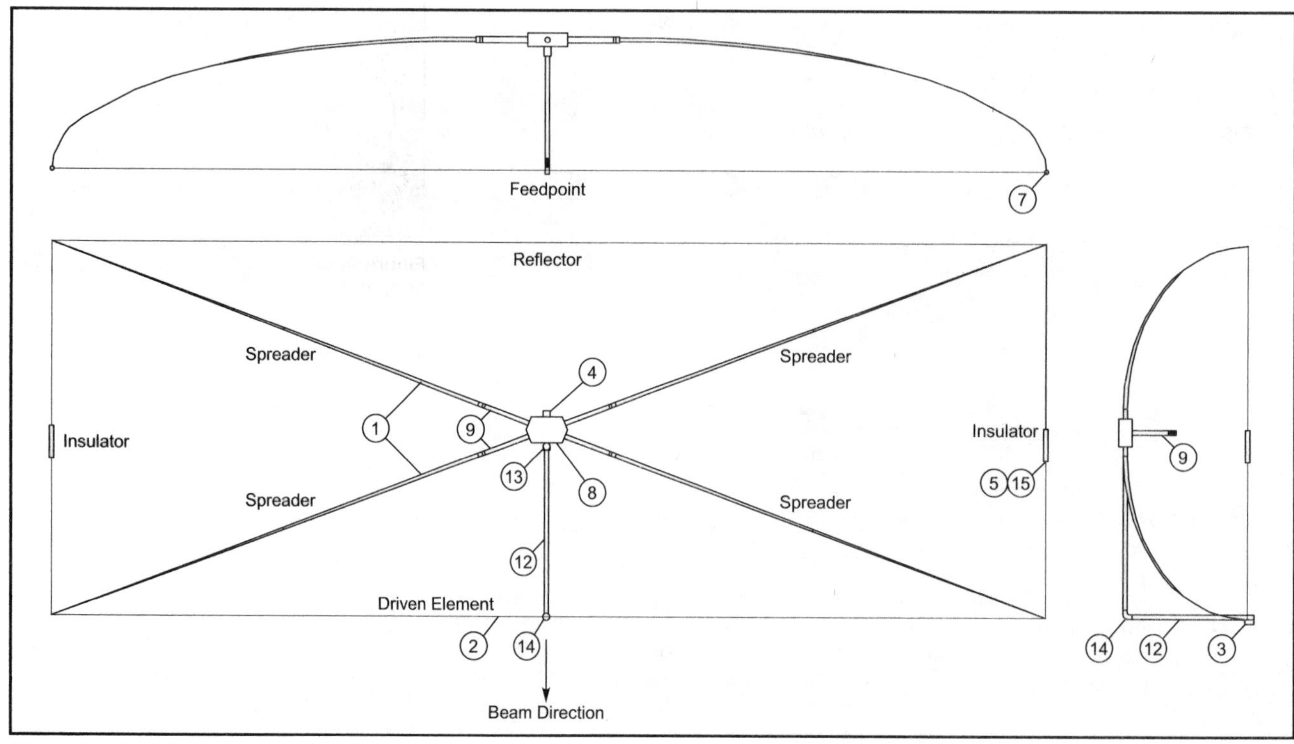


**Figure 2—Overview of the antenna and its components. Circled numbers refer to materials in Table 1. The mast and mast support bracket are not shown. Details are available at www.arrl.org/files/qst-binaries/black-widow.zip.**

**Figure 3—The center hub (left). Note the glued-in (epoxy) threaded sockets for the spreaders. The completed hub (right) shows the mounting stub extending upward and the feed line support extending to the lower left.**

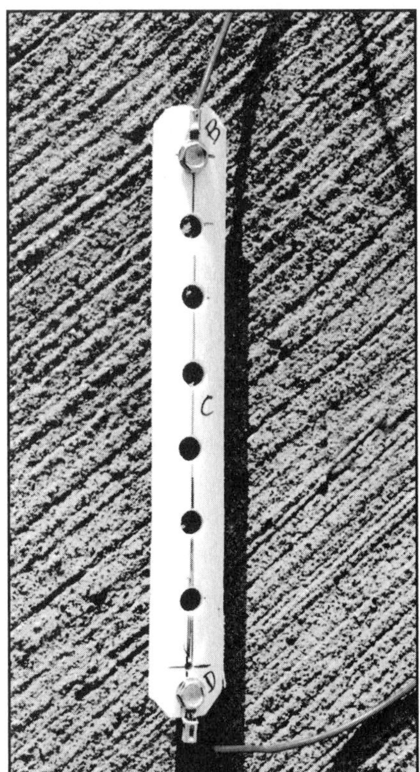


Figure 4—One of the two insulators for the wire elements.

Figure 5—The mounted terminal strip on the feeder arm.

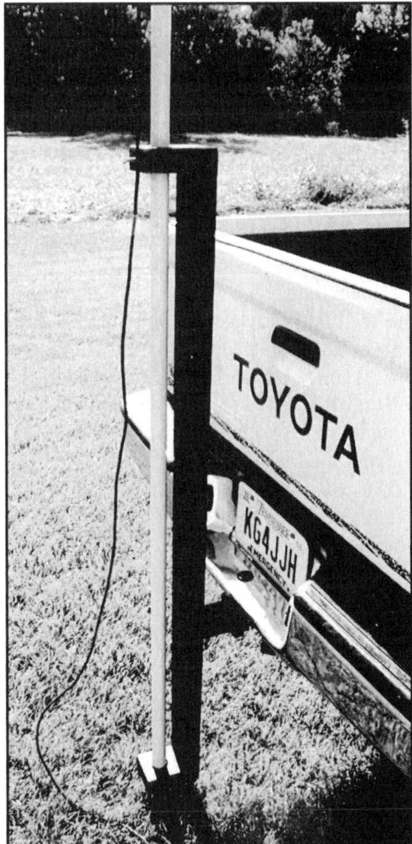


Figure 6—A view of the antenna mast support bracket.

## Spreaders

Modify each of the four fishing poles as follows. Unscrew the handle cap, roughen two inches of the inside surface with sandpaper, and epoxy a 12″ wooden extension pole (threaded on both ends) into the fishing pole handle. The original pole is too flexible and does not exert enough tension to keep the wires tight, so a portion of the small end must be removed. Extend the pole to its full length, and trim the small end so that the total length from handle to tip (including the wooden pole) is 106.5″. Use a hacksaw or band saw to keep the fiberglass from cracking. Epoxy a screw hook into the end of the pole. (Note: If the screw hook diameter is too large to fit into the end of the pole, cut another 0.5″ to 1″ off of the pole until it fits. Make all four poles the same length.) Spray paint the bare wood with Krylon *Ultra-Flat* black spray paint, and when dry apply two coats of clear lacquer.

## Center Hub

The center hub is made from 0.5″ thick poplar wood to minimize weight. Cut the pieces on a table saw and glue them together with yellow wood glue. Cut two of the plastic extension-pole handle sockets in half so that two threaded sockets are produced from each. Drill 1″ diameter holes in the hub, and epoxy the sockets into each hole. I chose to permanently attach the mast mount (which is a wooden extension pole threaded on one end) by drilling a 0.875″ hole through the top and bottom and securing it with epoxy. Extend the non-threaded end through the hub and trim off the excess after the glue sets.

Drill a 0.625″ hole for mounting the SO-239 connector on one side of the hub. Drill and mount the CPVC drop ear elbow on the opposite side, and drill a 0.3125″ hole through the drop ear elbow into the hub. Sand and paint the hub the same as the fishing pole handles. Figure 3 shows a completed hub.

## Insulators

The ends of the antenna elements (within dimension C of Figure 1) are supported by a pair of insulators that maintain a fixed distance between the wire ends. The insulators are made from 6″ × 0.75″ pieces of scrap 1¼″ (nominal) PVC pipe cut lengthwise. Six 0.3125″ holes are drilled into the PVC insulators to reduce weight. Drill a small hole for a #6 thread-cutting screw to attach the element ends to the insulators. Include the diameter of the solder rings when spacing the holes in the insulators. A view of each of the insulators can be seen in Figure 4.

## Wire

Cut wires to the following dimensions (including 0.4″ for each loop of wire around a screw hook):

Two each, ½ Driven Element = ½ A + B + 1 Loop

= 97.315″ + 29.875″ + 0.4″ = 127.59″ or 127.625″

One each, Reflector Element = A + 2D + 2 Loops

= 194.625″ + 72″ + 0.8″ = 267.425″ or 267.5″

Mark the corners (Figure 1, dimension intersections A-B and A-D) on the wires using a permanent marker. Strip ½″ insulation from the wire ends at the feed point for the driven element. Attach solder rings at the ends that will attach to the insulators.

## Feed Point and Feed Line Support

Coax is carried to the feed point of the driven element from an SO-239 connector on the hub, through the hub, and through an “L” made of CPVC pipe. The horizontal pipe length is 31.5″, and the vertical length is 24″. These lengths were determined with the antenna wires installed and the terminal block floating from its vertical support. To account for construction differences, cut the pipe pieces slightly long and trim them in small increments.

Connect the coax to the SO-239 connector first, then feed the coax through the hub, the drop ear elbow, the horizontal pipe, the 90° elbow, the vertical pipe, and out of a hole drilled in the pipe just above where the terminal block will be mounted.

Cut two terminals free from the strip. Clamp the driven element's feed point wire ends into each side of the terminal block. Then loop the wire around the open hooks at the corners using the permanent marker lines as an aid. Attach the wires to the insulators. Dry fit all pipe pieces until the terminal strip can be mounted to the lower end of the vertical pipe such that the driven element is straight and parallel to the reflector. Attach the two-terminal block with a small sheet metal screw. Figure 5 shows the mounted terminal strip.

Strip the coax and attach its shield and center conductor to the terminal strip. When complete, check all wire dimensions with a measuring tape and make adjustments as necessary within an accuracy of $^1/_8$". Remark the position of the screw-eye mounting loops, if necessary. Align and glue the CPVC parts. (If you prefer not to glue the CPVC, leave enough slack in the coax to allow removal of the pipe and terminal block assembly, and attach the pipes and couplings with screws.) Paint the feed line support to match the hub.

## Mast and Support

The painter pole can be supported in a variety of ways but should be guyed at three points. I fashioned my support from 2 × 4s to fit my trailer hitch and used the pickup truck tie points and a ground stake as guying anchors. The bed of the truck also serves as a convenient platform to raise and lower the mast. A PVC coupling, with holes drilled for twine guy lines slips over the pole. Be sure to drill and install a small screw into both ends of the plastic handle socket that connects the painter pole to the hub. This pins the antenna and prevents it from turning on the mast.

The finished Black Widow weighs less than 5 pounds and is stable at a height of 15 feet in windy conditions. On a calm day it could be extended to the full 23 feet, improving the gain and lowering the angle of radiation. Figure 6 is a view of the antenna support bracket and Figure 7 shows the erected antenna.

## Setup

Figure 8A shows the Black Widow ready for assembly. Remove the rubber stoppers from the ends of the fishing poles and extend them to their full length. Screw all four fishing poles into the hub. With the assembly upside down on the ground, attach the driven element wires to the feed point and wind the wires one loop around the screw hooks at the marks on the wire. As you progress, the fishing poles will begin to flex upward. Place the PVC pipe coupling with attached guy wires (nylon twine) over the top of the

**Figure 7— The erected antenna. In the air, the antenna resembles a black spider with its legs hanging down.**

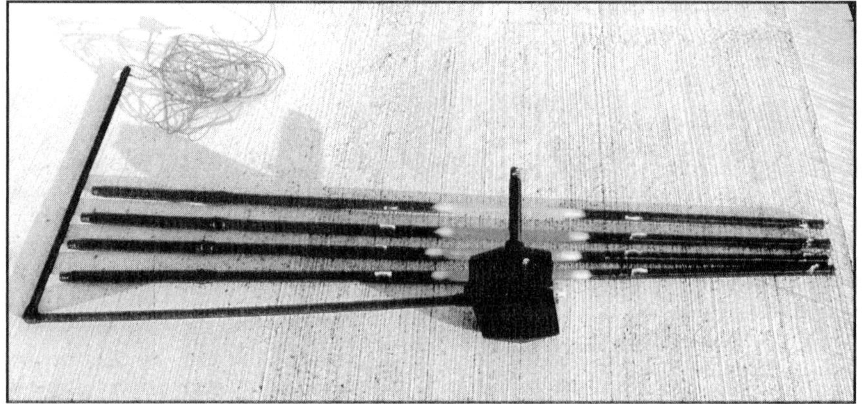

(A)

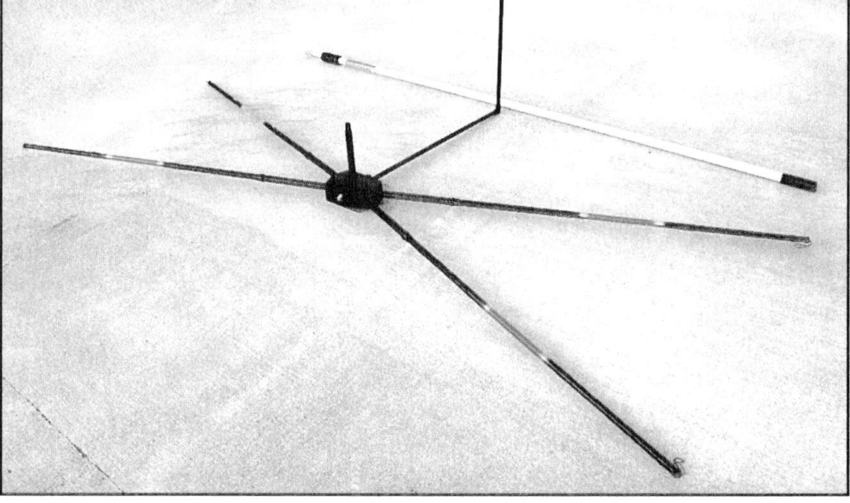

(B)

**Figure 8— (A) Broken down into its major components—spreaders, hub, feed line support and wire elements, the Black Widow is ready for transport. The partially assembled antenna (B).**

painter pole. Pick up the antenna, flip it over and screw the painter pole onto the mast. Secure the antenna by inserting a screw into the socket on both ends.

Connect a feed line to the SO-239 and secure it to the mast with Velcro or electrical tape. To help suppress RF currents on the outside of the coax shield, wind five turns of coax in a 6 inch coil near the antenna.

Place the antenna and painter pole mast into a suitable support. To prevent the painter pole from bending, raise the middle section first and adjust the guy wires as needed. Lower the middle section and raise and tighten the top section. Now raise the middle section again and tighten. Clothesline tighteners were used on all guy wires to allow adjustment without having to re-tie the twine.

## Results

The antenna was designed for PSK use at 21.070 MHz, but tests conducted with an MFJ antenna analyzer at heights of 15 and 23 feet reveal a fairly flat response of 1.2:1 to 1.3:1 across the entire 15-meter band. The *EZNEC* models predicted the antenna to have a gain of 9 dBi at the 15 foot level and 10.45 dBi at the 23 foot level. Likewise, the predicted SWR is 1.11:1 at 21.1 MHz.

Upon completion, I wanted to see if the Black Widow would perform as I had hoped. Tropical storm Isadore had just passed through our area, so it was clear but very windy. With an eye out for the wind, I mounted the antenna on the painter pole at a height of only 15 feet.

A good test for any antenna is to see how well it performs at QRP power levels. With 5 W from my FT-817, my first PSK contact was a station in San Diego, who gave me a signal report of 579. Not bad, considering that distance from my QTH in Tennessee is 1900 miles. Next, I swung the beam toward Europe and contacted a station in the UK who gave me an RST of 589 and an uninterrupted QSO for about 15 minutes. Wow, I was getting excited—that's over 4000 miles!

With the assistance of my brother (NG4T), further SSB/CW tests were conducted using his TS-2000 at power levels up to 100 W with great results. The Black Widow is easy to build, transport and set up. Figure 8A shows the antenna ready for transport and Figure 8B shows it partially assembled. It will outperform many portable 15 meter antennas. Good luck and have fun!

### Notes

[1]**www.cebik.com/moxpage.html**.
[2]**www.qsl.net/ac6la/moxgen.html**.
[3]**www.eznec.com**.

*Allen Baker, KG4JJH, received his license in September 2000, after a lifelong dream of becoming a ham. He holds a BS in Industrial Engineering from Tennessee Technological University and works as an instrument and controls engineer with the Department of Energy in Oak Ridge, Tennessee. Allen is active on the digital modes (10 through 40 meters) and loves to experiment with antenna designs. He can be reached at* **kg4jjh@arrl.net**.

By Al Alvareztorres, AA1DO

# Two on 10

Hankerin' for more performance on 10 meters? Wanna greet the upcoming sunspot peak with gusto? This home-brew two-element beam is the perfect introduction to rolling your own gain antenna.

Living in a condo has many advantages, none of which is being able to mount a tribander on a 60-foot tower. So I make do with a long, thin random wire that works nicely as long as the New England wind, snow and ice don't conspire to give my hamming a holiday (which happens more often than I'd like). And although it's somewhat directional on the higher bands, I haven't figured out how to rotate 200 feet of wire without the neighbors becoming suspicious. One answer is to operate mobile. A bumper-mounted vertical is fine for casual operation, but it leaves a lot to be desired when mountaintopping for rare DX. With the solar maximum just around the corner, I decided that a portable 10-meter beam was necessary.

The beam had to fit in the trunk of my Subaru (limiting the largest component to about four feet in length) and had to be easy to assemble and erect on site by one person. In this article I'll describe the antenna and provide some construction tips that may help you avoid some pitfalls if you take on this worthwhile project.

### The Boom

From past experience I know that TV masts make good booms for smaller antennas. They're lightweight, strong and readily available at most RadioShack and home stores. The light-duty stuff is plenty strong and comes in five-foot lengths. That was my starting point.

At 28.4 MHz, for an antenna made of tubing and not supported at the ends, a half wavelength is 491.8 divided by 28.4 MHz, or 17.3 feet. To accommodate my "Subaru factor," four feet divided by 17.3 feet produces a boom length of 0.116 wavelength, a size that gives a nice gain and a feed-point impedance that can be easily matched to your coax line.

The TV mast (with the crimped end lopped off) fits in my trunk and allows two elements to be mounted 4 feet apart and fed with RG-58 coax. So far so good. I would be building a two-element beam.

Now, how to mount the elements to the boom and the boom to the mast (another 5-foot TV mast section)? In the past I had used a **U**-bolt and clamp arrangement, but this technique requires care in keeping the elements parallel to each other and to the ground. This is fine for permanent installations, but not something to be bothered with while operating portable.

I decided on right-angle pieces permanently mounted to the boom (see Figure 1). I used ½-inch 1 × 2 aluminum angle scrap because it "looked about right." Your local hardware store has aluminum angle in various dimensions and lengths. I cut six, 3-inch pieces of angle to make the **U**-bolt mounting brackets—two to hold each element and two to hold the mast.

Drilling the two holes in the angle's smaller dimension—the part that attaches them to the boom—isn't critical as long as you drill the holes on the boom the same distance apart. The angles will be permanently mounted to the boom using 2-inch bolts, nuts and lock washers. When you mount them, be careful not to crush the tubing. It's not terribly strong, but it *is* lightweight. We're going for portability here!

The holes in the larger dimensions should be tailored to allow the mounting of the 1¼-inch **U**-bolts for the elements and the 1¾-inch **U**-bolts for the mast. Because the element mounts must be as parallel as possible and the boom mount must be at right angles to them for maximum efficiency (and so your antenna doesn't look like it's under the influence), make the boom holes with a drill press if possible.

### Now for the Elements

Most beam antennas are made with aluminum tubing because it's strong, lightweight and available in sizes that "telescope" into each other. The telescoping feature is important. It helps in transportation and makes tuning the antenna a snap.

The beam's driven element should be 17.3 feet. The length of a reflector for a two-element Yagi with 0.116-wavelength element spacing should be 18 feet ½-inches.

I needed 35 feet of tubing (plus some to fit inside each telescoping joint for support). Because the tubing comes in 8-foot lengths, this worked out to five lengths of assorted sizes. The three telescoping sizes available at my local hardware store were 1 inch, ⅞ inch and ¾ inch—perfect! Because the 1-inch section was going to be the center part of the two elements, I picked up **U**-bolts and nuts while I was there. You'll also need eight hose clamps sized to fit your tubing.

This is how the material was cut up. One 1-inch tube was cut in half, yielding two four-foot lengths. The two ⅞-inch tubes

**Figure 1—Two pieces of angle aluminum are attached to the boom with two nuts, bolts and lock washers. Your drilling must be accurate, so use a drill press if possible. In addition to drilling the boom holes, you'll need to drill four holes in each angle piece: two for accepting the boom-mounting nuts and two for the U bolts that will hold the element.**

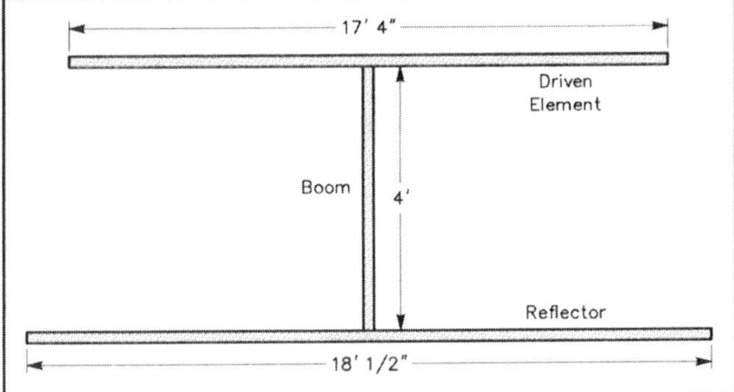

Figure 2—Dimensions and mounting configurations for the boom, driven element and reflector element.

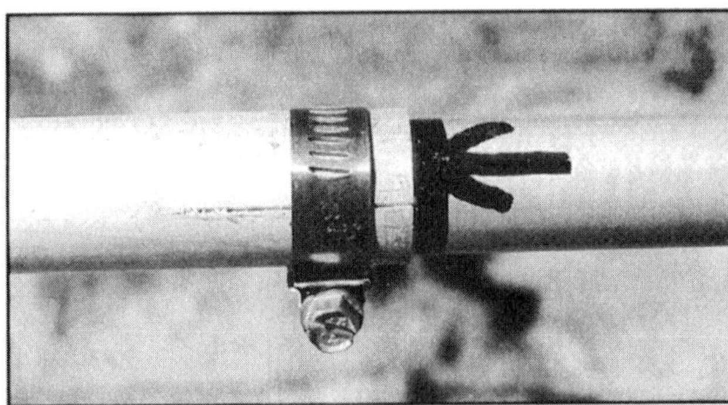
Figure 3—Use hose clamps to compress the slotted ends of the telescoping elements and keep everything in place. Marking the exact positions of the sliding tubes makes it much easier to assemble the antenna in the field.

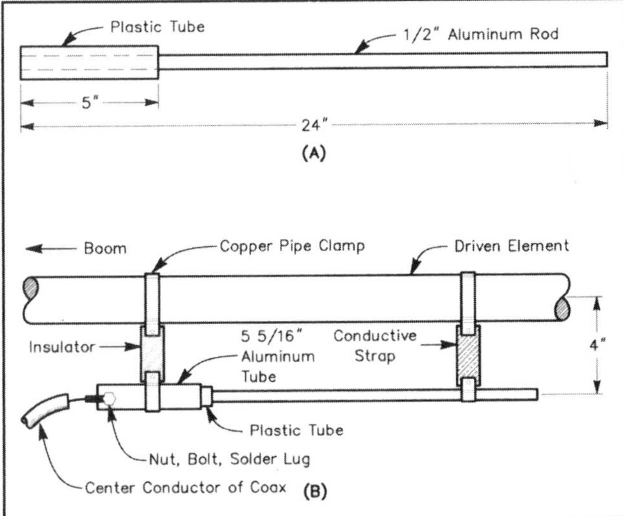

Figure 4—Making your own gamma match is easier than you think. Construct the variable capacitor element by sliding a 5-inch piece of 1/2-inch diameter plastic tubing over a 24-inch piece of 1/2-inch diameter aluminum rod (A). Then, slide the plastic-sleeved end of the rod into a 5 5/16-inch long, 3/4-inch diameter aluminum tube and attach the entire assembly to the driven element (near the boom) using one plastic insulator and one conductive strap as shown (B). Note that the assembly must be separated from the driven element by 4 inches, center-to-center.

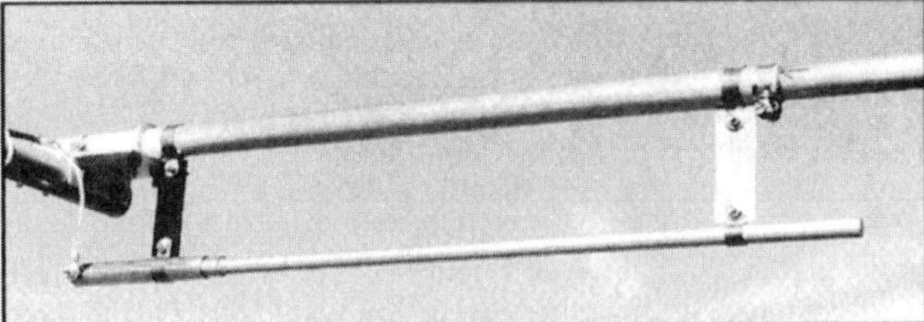
Figure 5—The finished gamma match mounted and ready for action.

were cut in half to yield four four-foot lengths. One 3/4-inch tube was cut in half to yield two four-foot lengths. From the remaining 3/4-inch tube I cut off an 8-inch piece (for later use in the gamma match) and cut the remaining length in half to yield two lengths a little over 3.5 feet each. I then took the 1-inch tube and cut a slot in each end to a length of about 1 1/2 inches. Pushing the tube end-wise into a band saw makes a really nice double-slot arrangement. I did the same at one end of each 7/8-inch tube. When the elements are assembled, hose clamps will pinch the slots closed and keep the element sections in place (see Figure 3).

## Slappin' Together Time

In the garage I erected a 3-foot tripod. I then cut a small piece off the un-swaged end of the second five-foot TV mast (so it would fit in the trunk) and installed it into the tripod. I mounted the boom on the mast with **U**-bolts and clamps and attached the two 1-inch tube sections to each end of the boom with **U**-bolts and centered them for balance. I then slid the unslotted ends of the four 7/8-inch tubes into the ends of the 1-inch tubes, holding them in place with hose clamps. I inserted unslotted ends of the four remaining tube sections in place (using the two shorter 3/4-inch tubes on the driven element). You won't believe how big a 10-meter beam seems when it's inside a garage!

## Some Last Element Details

I drilled a hole at the center of the driven element and installed a bolt to attach the shield of the coax. I drew a ring around both 1-inch tubes with permanent markers to show the exact center for easy assembly. I used black when marking the driven element and red on the reflector. That way, in the field I wouldn't have to stop to figure out what was what (that's also why I cut slots into only one end of some of the element sections).

## Feeding the Antenna

As you may have noticed, this antenna uses "plumber's delight" construction. The driven element isn't split into two legs like a conventional dipole. In this case, the driven element is one piece, and everything is shorted to the boom, to the mast and to ground. To top it off, the whole mess is fed with unbalanced coax. So, how does it work? Like magic! And the magic words are *gamma match*. There are actually several ways to feed a plumber's delight antenna, but the gamma match is probably the simplest.

How a gamma match works is beyond the scope of this article. In short, the braid of the coax is connected to the center of the driven element (since this is where the voltage null occurs in a half-wave conductor). The center conductor of the coax is connected to the same driven element through a capacitor some distance away from the center. In the old days we used tuning capacitors from discarded AM radios. Tuning caps are as scarce as hen's teeth nowadays, so I decided to try a technique I'd come across in the 1974 *ARRL Antenna Book*—incorporating the capacitor into the structure of the gamma match.

## Building the Gamma Match

I took the 8-inch piece of 3/4-inch tubing that I had set aside before and cut it to 5 5/16 inches. I cut a piece of 5/8-inch plastic

**The shield braid of your coaxial cable attaches to the driven element using a nut, bolt and solder lug. The center conductor, however, must attach to the end of the gamma match capacitor, as shown here.**

### Bill of Materials

(2) 5-foot light-duty TV mast
(1) 1-inch x 8-foot aluminum tube
(2) $^7/_8$-inch x 8-foot aluminum tube
(2) $^3/_4$-inch x 8-foot aluminum tube
(1) $^1/_2$-inch x 4-foot aluminum rod
(1) 1-foot section of clear vinyl tubing
(1) 2-foot aluminum angle
(2) 1$^3/_4$-inch U-bolts
(4) 1$^1/_4$-inch U-bolts
(8) Hose clamps to fit on 1-inch tubing
(6) 2-inch bolts & hardware
(2) 1$^1/_2$-inch bolts & hardware
(4) $^1/_2$-inch bolts & hardware
(1) 3-foot tripod
(3) 1-foot metal tent pegs

tubing to a length of 5 inches and cut a $^1/_2$-inch aluminum rod (tubing will work) to 24 inches. Sliding the plastic tubing onto the $^1/_2$-inch rod until their ends were flush, I now slid this assembly into the 5$^5/_{16}$-inch tube until $^1/_2$ inch of the plastic tube was left exposed (see Figures 4 and 5). I now had a capacitor!

I drilled a hole near the end of the 5$^5/_{16}$-inch tube and installed a small bolt for the center conductor of the coax. This assembly was mounted to the driven element so that the larger end (the one with the bolt) was directly under the center of the element and the two tubes were four inches apart center-to-center.

The gamma match is held on by an insulated strap at the end closer to the center of the driven element and by a conductive aluminum strap at the other end. The locations of the straps aren't critical at this point. The straps themselves can be made of any sturdy insulating and conducting materials. I used flat plastic stock and flat aluminum stock (1 inch by $^1/_{16}$ inch worked fine) held in place by copper clamps (designed to hold copper pipes to a wall). These clamps come in all sizes, are easily bent to the proper size, already have holes in them for attaching to the straps and are inexpensive. Mechanically, everything looked good! But would it work?

### Tuning the Antenna

I was fortunate that it was a beautiful summer day and that I had my wife, Donna, AA1DQ, to help me. I disassembled the monster in the garage and reassembled it on the lawn. Everything went together nicely in about 15 minutes. I attached the braid of the coax to the driven element and the center conductor to the gamma match. The fact that it was only four feet off the ground would have little effect on the tuning, although the overall performance would be affected by the high angle of radiation. Leaving the hose clamps over the element slots loose, I adjusted the driven element length to about 17 feet and the reflector to about 18 feet $^1/_2$ inch.

I would make the adjustments while Donna, visible through the shack window, keyed the transmitter and recorded the SWR readings. It goes without saying that visual (or some other positive) contact is imperative for safety. She could see that I was clear of the antenna before keying the transmitter.

I find that it's best to keep a written record when tuning an antenna (even if it's only a dipole) so that I know where I am and which way I'm going. I make a chart with frequency on the Y axis and antenna length on the X axis. I then enter the lowest SWR point (resonance) at the appropriate X-Y position. As I change the length I can easily see what's happening.

If you find that the SWR at your chosen frequency is unacceptable, begin adjusting the gamma match by sliding the center bar in or out. If you can't achieve a match, slide the entire matching section toward or away from the center of the driven element. As a last resort, adjust the driven element length. This will also have an effect. Remember to keep records. Otherwise you may get your adjustments all out of whack and won't know where you are. When you're done, tighten the hardware on the gamma match, as it will not be moved again.

I was lucky. After only a few adjustments I obtained a 1:1 match at any chosen frequency (28.0 to 28.5 MHz). A match of 1.3:1 was attainable beyond these frequencies (up to 28.6 MHz). Your mileage may vary.

I was overjoyed. The beam showed very good side rejection and a respectable front-to-back ratio. I marked the element sections at their contact points with a ring using the same permanent markers. When erecting the system I could simply slide the sections to the rings and tighten the hose clamps. There was nothing left to do but try it out on Beseck Mountain. Along the way I got some foot-long metal tent pegs to hold the tripod steady.

### The Verdict

The antenna has been used several times mountaintopping and contesting. It performs well and can be erected by one person in about 15 minutes. It was well worth the effort. I have since gotten another section of mast and, with two people, it can easily be put up at 10 feet.

I haven't experimented with the reflector length yet to see the effect on the gain and the front-to-back ratio. As they say, "If it ain't broke, don't fix it!"

**225 Main St, Newington, CT 06111; aa1do@arrl.org** QST

# A Portable Two-Element 6 Meter Quad Antenna

**Assemble or knock down this antenna in less than 5 minutes. Who says Field Day is only about HF?**

**Pete Rimmel, N8PR**

Constructed from PVC tubing, this portable 6 meter antenna disassembles easily for storage, and is easy to assemble at an RV campground or Field Day site. I devised a unique way to lock the spreaders into the center PVC crosses using the antenna wires of the quad. This key feature allows me to tighten the wires so that they won't sag while in use, but lets me loosen them for disassembly and storage in a small space such as in my RV.

## Construction

You can construct this quad from parts available at any home improvement store. The PVC dimensions are not critical, and can be within ¼ inch. The center crosses of ¾-inch pipe are compatible with 1-inch Ts that were split to create the boom connections. The inner diameter of the 1-inch T matches the outer diameter of the ¾-inch crosses with a bit of trimming or filing. This larger size PVC is more rigid and makes the antenna more stable.

I used #14 AWG stranded insulated wire for the antenna elements. For me, insulated wire is less troublesome to store in my RV, and won't snag or break while in storage.

## Gathering Materials

Cut four 40-inch pieces of schedule 20 × ¾-inch PVC for the driven element spreaders, and four 41¼-inch pieces for the reflector spreaders. From a 10-foot section of schedule 20 × 1 inch PVC, cut two 11-inch pieces for the boom. Use the rest for the vertical mast.

You will also need two ¾-inch PVC crosses (these have four openings); two 1-inch PVC Ts (these have three openings, see Figure 1); one 1-inch PVC cross; and six 1¾-inch diameter stainless steel hose clamps. Cut an 18-foot, 10½-inch length of insulated #14 AWG stranded wire for the driven element and a 19-foot, 5¾-inch length for the reflector. Figure 2 shows the antenna disassembled, with the parts ready for storage.

You'll also need three 2-inch long pieces of ⅜-inch heat shrink tubing, coax sealant, black electrical tape, Velcro® or similar hook-and-loop fastener strips, and a length of 52 Ω coax to reach to your station.

## Preparing the Boom

Cut two pieces of the 1-inch PVC pipe to 11 inches long. Place the 11-inch long 1-inch diameter PVC tubes into the 1-inch cross which also connects to the boom, as seen in the lead photo. The rest of the 1-inch PVC goes into the lower part of the T to form the mast.

Make a pair of "five opening fixtures" for connecting the four cross pieces to each end of the boom as follows. Make two "half-Ts" by cutting the long side of

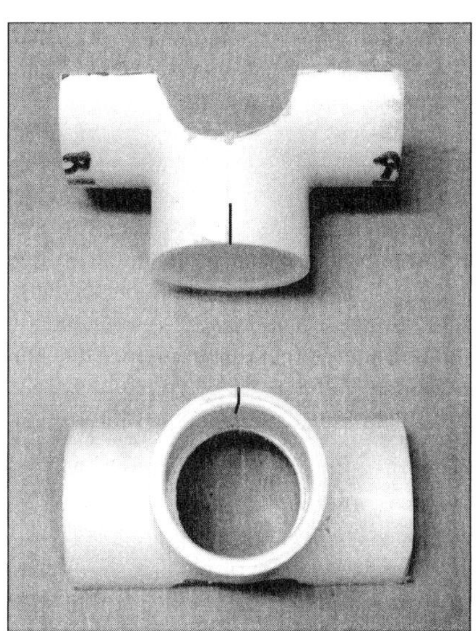

**Figure 1** — Split a PVC T and file away material so that with a four-way PVC cross you have a fixture with five openings.

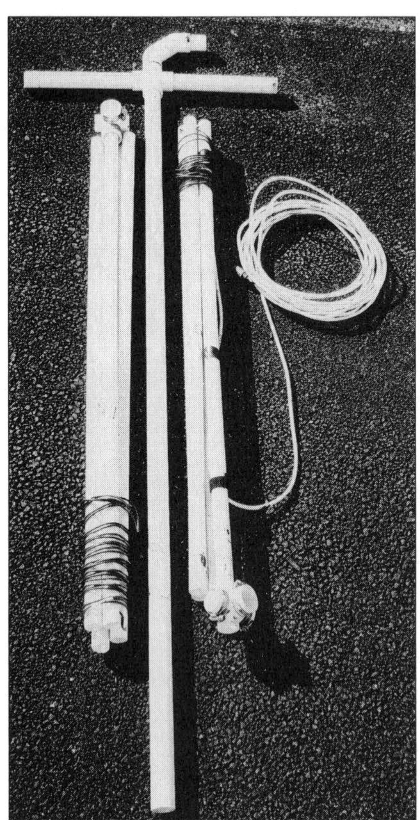

Figure 2 — Easy-to-stow disassembled antenna components.

a 1-inch T with a hacksaw as in Figure 1. Cut a slit in the round part of this fitting. Shape the cut area, and remove a half-circle of material using a rasp or file so that it will fit snugly over the ¾ inch cross. Use two hose clamps to fit each half-T to each PVC cross (see Figure 3). The hose clamps should pull the T snug to the smaller cross without deforming it. A third hose clamp compresses the slit and holds the element on the boom.

### Making the Spreaders

Drill ⁵⁄₆₄-inch holes ½ inch from the end of each of the eight spreader pieces, going through both walls of the pipe. Take care to center these holes so the wire will go straight through the center of the spreader. Cut J-shaped openings in three of the 40-inch long spreaders and all four of the 41¼-inch spreaders. Following the progression of Figure 4 from left to right, drill a second hole in each spreader at about a 30° rotation and about 1¼ inches in from the end. Again, be sure to drill straight through both walls of the pipe. Draw some guidelines on each of the seven pieces of PVC. Drill a series of holes to form an upside down J on both sides of each tube. Be sure to leave PVC in the gap between the holes on the right and the single hole on the left so that the wire can lock in two positions. Carefully "connect the dots" with a drill, and remove all the material between the holes. This forms the unique adjustable J feature of the wire guide on seven spreaders.

### Assemble the Reflector

Strip 1 inch of insulation from each end of the longer #14 AWG wire (19 ft 5¾ in) and twist the stranded wire to keep it neat. Thread the wire through the eight J holes of the four 41¼-inch spreaders. Place a 2-inch piece of heat shrink tubing over one end of the wire. Overlap the two ends by 1 inch, twist and solder them together. Slide the heat shrink over the soldered connection and heat it to waterproof the connection. You now have a loop with the wire going through the four outer ends of the spreaders. Position the wire in the inner part of the J on all four spreaders.

Insert the four spreaders into one of the two ¾-inch crosses. After you seat the pieces snugly into the crosses, move the wire to the outer part of the J to tension the wire element. You might not need to tension the wire on all four spreaders. Twist the pipes so that the holes line up and the wire lies in a flat loop when assembled. Your reflector is now completed.

### Assemble the Driven Element

Strip 1 inch of insulation from each end of the insulated 18 foot, 10½-inch long #14 AWG wire. Prepare a length of 52 Ω coax (RG8-X, RG-58, or RG 213) by stripping the outer jacket back about 1½ inches. Comb the braid out so that it can be separated from the center conductor and twist the braid into a single strand. Cut it to a 1-inch length. Strip ¾ inch of the insulation away from the center conductor. These dimensions contribute to the overall length of the driven element.

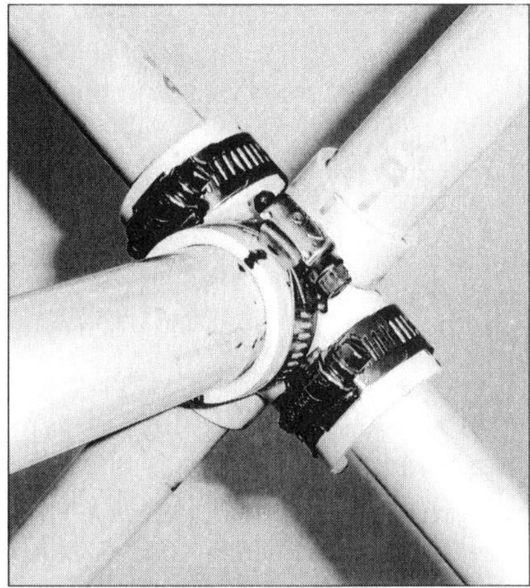

Figure 3 — The five-opening fixture ties the boom end to the four spreaders.

Thread the insulated #14 AWG wire through the holes in the four 40-inch spreaders. Be sure that the bare ends of the wire are near the spreader without the J cut into the end. This is where you connect the coax. Slide a 2-inch piece of shrink tube over each end of the #14 AWG wire. Carefully wrap the center conductor of the coax to one stripped end of wire and solder it. Do not overheat. Wrap the other wire around the braid of the coax and solder it in place. Slide the two pieces of shrink wrap over the soldered connections and heat them. Seal the remaining area of the coax braid with a small amount of coax sealant and cover with black tape to completely waterproof the connection (see Figure 5). Use some loops of Velcro (or black tape) to secure the coax to the PVC tubes. Assemble the driven element by placing the four spreaders into a ¾-inch cross, and place the wires in the J cuts to tension the wire and rotate so that the loop is flat.

### Assembling the Antenna in the Field

It is easy to field-assemble and disassemble the antenna; see also the *QST* in Depth web

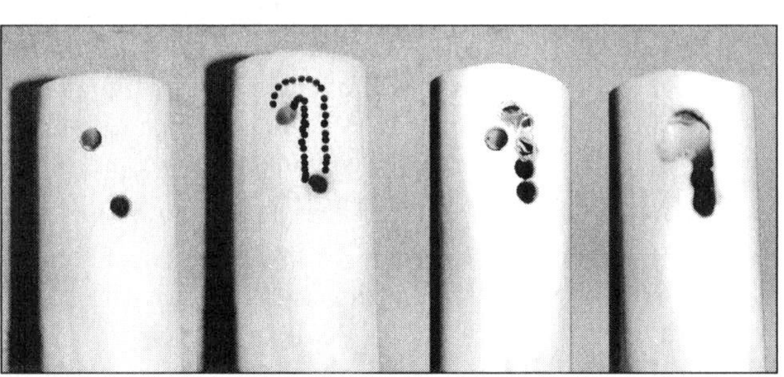

Figure 4 — Steps for making the J holes.

page.[1] Separate the two elements and lay the reflector parts on the ground. Be sure that the wire is in the long part of the J holes. Place the four elements into the cross. Move the wires into the shorter part of the J holes to tension the wire and keep the spreaders from dropping out of the cross. Slide the director onto the boom and partially tighten the hose clamp over the slit in the "half-T" to secure the element to the boom. Assemble the driven element as you did the reflector and mount it onto the boom using another hose clamp. Be sure to orient the elements in the polarization you want before tightening the hose clamps completely. Route the coax along the element and boom to the mast using Velcro loops to keep it in place.

When you take the antenna apart, leave the cross piece attached to one of the four spreaders for storage. Wrap the element wires around the four spreaders to keep them in a neat package.

**Figure 5** — Detail of the coax feed line connection.

### Final Comments

You can easily choose the polarization you want. For vertical polarization, place the driven element so that the coax is on a horizontal spreader, and the antenna is in a "diamond" configuration. For horizontal polarization, place the diamond shape so that the coax is at the bottom of the element. You can also orient the elements for 45° polarization.

This antenna is designed to resonate at 50.1 MHz. The VSWR should be less than 1.5:1 from 50.0 to 50.6 MHz. The element spacing is designed to give a 50 Ω match.

[1]www.arrl.org/qst-in-depth

Photos by the author.

Amateur Extra class licensee Pete Rimmel, N8PR, was first licensed in 1960 as KN8UNP in Cleveland, Ohio. He is a Life Member of ARRL and QCWA, holds DXCC #1, 10 Band DXCC, and 5 Band WAZ. Pete is President of Local Chapter #69 of QCWA and served as officer and director of the South Florida DX Association. He earned a BSChE from the University of Cincinnati, and is a retired USCG Master of Passenger Vessels. He recently retired as an NFPA Certified Marine Chemist. Pete enjoys DXing and contesting on 160 to 2 meters, satellite operation, and EME. Other hobbies include sailing, scuba diving, and folk singing. You can reach Pete at **n8pr@bellsouth.net**.

**For updates to this article, see the *QST* Feedback page at www.arrl.org/feedback.**

# Take-Down Yagi for the 2 Meter Band

## This collapsible antenna is perfect for foxhunting and other applications.

**Jerry Clement, VE6AB**

My 2 meter take-down Yagi is based on the tape-measure Yagi designed by Joe Leggio, WB2HOL, and described in the 22nd edition of *The ARRL Antenna Book*.[1, 2] My lightweight take-down antenna meets Joe's specifications, and includes quick-disconnect couplers as a main feature. The disassembled Yagi fits in a 2 × 6 ×10 inch netbook pouch (see Figure 1). Construction requires just simple hand tools, although a small bench-mounted drill press might prove beneficial.

### Making the Yagi Parts

Gather the parts listed in Figure 2. For the director element T, use a hacksaw to cut the barbed sections off all three ends, leaving only the smooth sections [Additional detailed images are available on the *QST* in Depth web page.[3] — *Ed.*] With the T mounted in a drill press (or bench vice), drill out a ½-inch diameter hole down the length of the part of the T that will mount on the boom. Do not drill through the end of this T, for cosmetic reasons. Cut off the barbed ends of the T for the reflector element, and drill completely through the part of the T that will slide in place over the boom. Drill the T for the driven element to ½-inch diameter completely through, allowing it to be positioned on the center of the boom.

Cut off only the barbed section of the T portion that slides over the boom. With a razor knife, remove the remaining barbs by cutting sideways along the T until you can slide 1½-inch length of Schedule 40 PVC pipe over each side. These PVC element keepers will hold the driven element segments in place on the T.

Cut the nylon-tube spacer into three pieces measuring 2 inches, 2¼ inches, and 2½ inches. Insert the 2-inch piece into the director T opening. Insert the 2¼-inch piece completely through the driven element T opening, and insert the 2½-inch piece through the reflector T, allowing ¼ inch to protrude from the end that faces the driven element (Figure 3).

Screw together a coupler pair that will make up the quick-disconnect boom feature. With a center punch, mark the center of one end, centered between the four terminal cavities. Place the assembled coupler pair in your drill press or bench vice, and drill a 5/16-inch hole completely through the coupler. Repeat for the second coupler pair. The couplers should slip nicely over the fiberglass boom.

Fully extend the 12-foot tape measure and disconnect or cut it loose with serrated scissors. Cut off 35⅛ inches for the director element (make it slightly longer for dressing the ends). Cut off 41⅜ inches for the reflector, and 17¾ inches for each side of the driven element. For safety, round off the ends of each element with your serrated scissors, and then file until smooth. Be careful not to shorten the elements to less than their required length while trimming the ends.

### Assembling the Yagi

Fasten the 35⅛-inch-long director element to its T. Use two ¾-inch lengths of Schedule 40 PVC pipe (element keepers) to capture the centered element in place on the T. Fasten the 41⅜-inch reflector element to its corresponding T, centering it and keeping it in place with element keepers.

Drill a 3/32-inch hole in the mounting end of each driven element, then fasten each driven element segment in place on its T. This will make it easier to solder the coax connections, and the hairpin match to the elements. I removed the paint around the drilled hole and roughed up the tape measure surface with a grinding burr (using a Dremel® tool). Place the elements on the

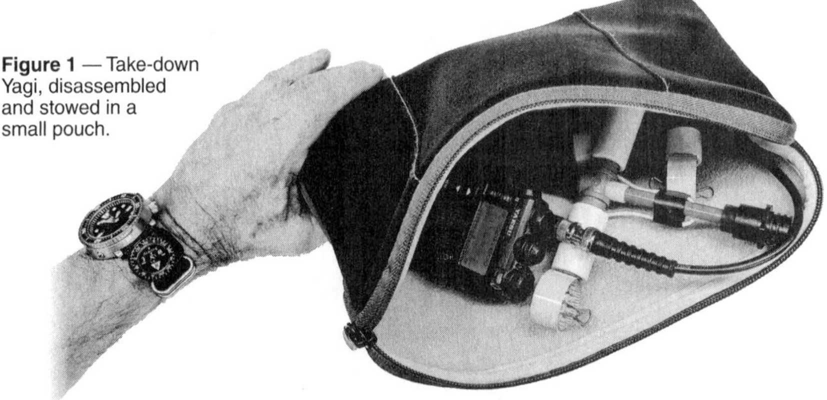

**Figure 1** — Take-down Yagi, disassembled and stowed in a small pouch.

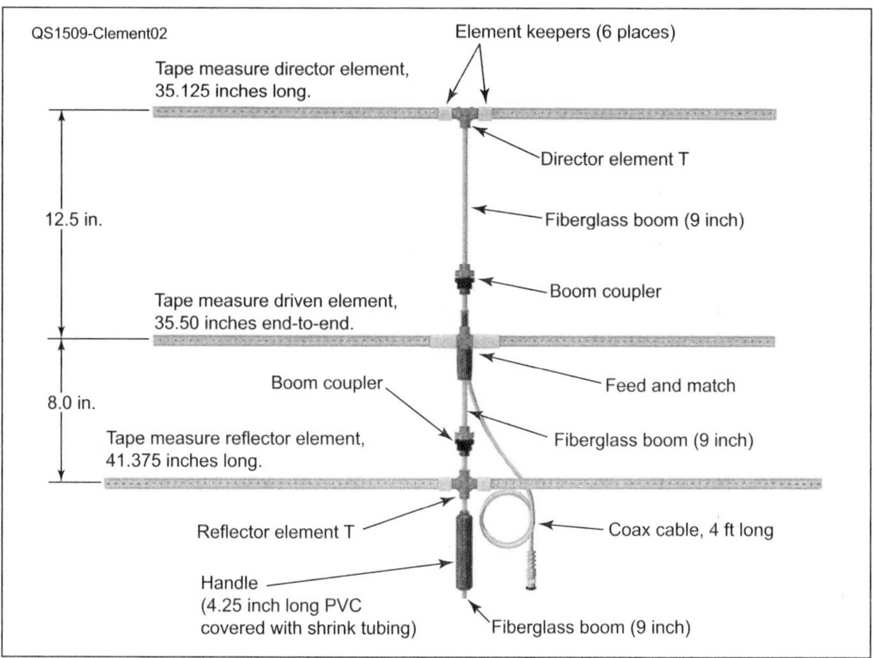


**Figure 2** — Take-down Yagi components and parts list.

Antenna elements — cut from 12-foot long, ½-inch-wide tape measure
Fiberglass boom — **www.homedepot.com/p/Blazer-International-Driveway-Marker-48-in-Round-Orange-Fiberglass-Rod-381ODM/202498049**
Element T — (3) Genova ½-inch diameter plastic coil fittings, modified (see text). **www.lowes.com/pd_22526-322-351405_4294822027__?productId=3455110**

Element keepers and handle — cut from 15-inch length of ½-inch Schedule 40 PVC pipe.
Element spacers — (see text) cut from 12 inches of ½ inch O.D. by ⁵⁄₁₆ inch I.D. nylon tube. **www.mcmaster.com/#standard-nylon-hollow-tubing/=tn1osa**
Couplers — modified from AMP 4 housings, two male #495197, and two female, #495091, **www.jameco.com/1/3/amp-circular-connectors-4-pin**

T, butted up against the ribs on either side of the T (¾-inch spacing), and then fasten in place with two 1¾-inch long element keepers (Figure 4). Drill through the holes in the elements where they sit against the T, to allow for easy positioning of the coax and hairpin match connections. Before soldering, we will assemble all the various components to the boom of the antenna.

Using the 27-inch length of the ⁵⁄₁₆-inch diameter fiberglass rod as a boom, slide the director assembly in place, and then the driven element assembly, and the reflector assembly. Next, align the individual components on the boom. I placed the six element spacers on a table beneath sides of each T for temporary support. Adjust the spacing between the center of the director element and the center of the driven element to 12½ inches. Adjust spacing between the center of the reflector element to the center of the driven element to 8 inches.

Make a pencil mark on the boom on either side of the element assemblies where they are positioned. Now slide them slightly out of position, and apply a very thin layer of 5-minute epoxy on the boom inside of the area where the three element assemblies will be positioned. Slide the element as-

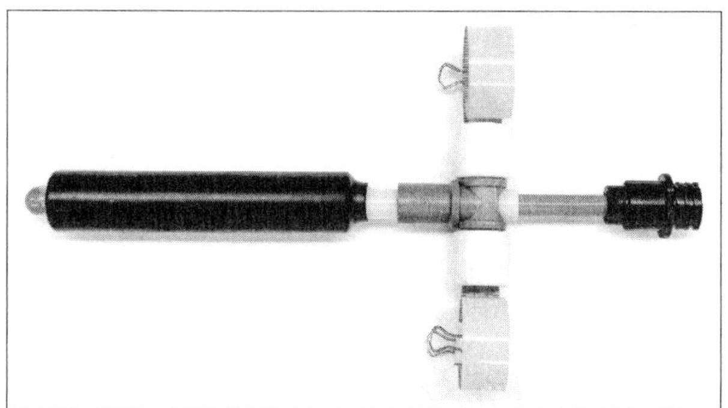

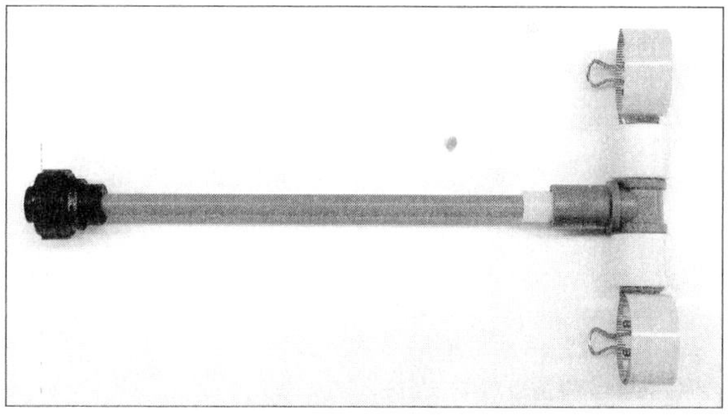

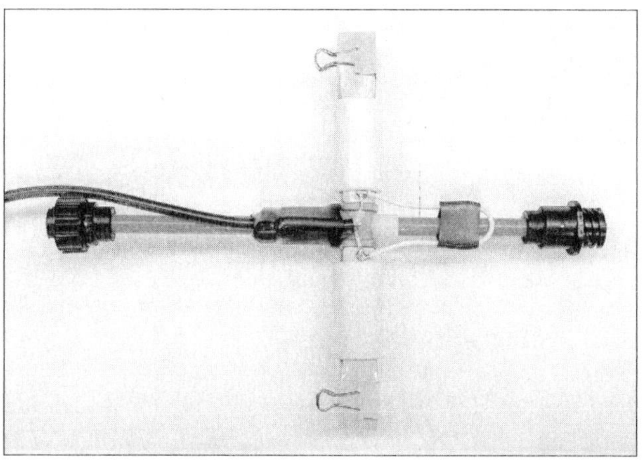

**Figure 3** — Disassembled antenna; handle and reflector element at left, driven element at bottom, and director element on right.

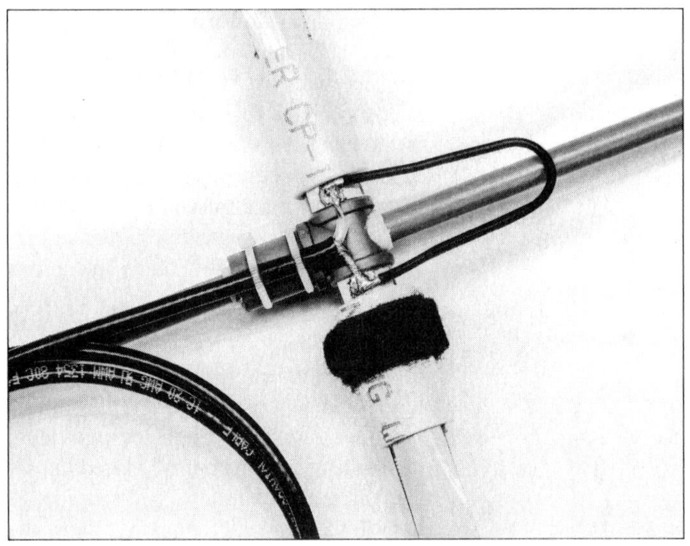

**Figure 4** — Feed point detail shows the center T with PVC element keepers, and the coax and hairpin match soldered to the driven element halves.

semblies back in place, making sure they are in alignment with one another. Allow the epoxy to set. Very little epoxy is required, as once the antenna is complete, we will pin the assemblies to the boom, ensuring that the components are securely fastened.

After the epoxy has set, install the couplers in place on the boom. Start by dividing and marking the boom into three equal 9-inch lengths. With a hacksaw, cut through the boom at the two marked locations.

Install one coupler at a time. Slide each mating half of the coupler in place on the boom. Square the antenna off once again, making sure the elements are aligned with one another. With each half of the coupler slid back from the cut in the boom, apply a very thin layer of 5-minute epoxy. Bring the coupler halves together, ensuring that the ends of the boom that you cut are in the proper position and butted up against each other. Then lock the two halves of the coupler together. Once more check the boom alignment, and allow the epoxy to set. Repeat these steps for the second coupler.

When the epoxy has set, pin all the assemblies in place to the boom. Use tiny ½-inch finishing nails as pins. Drill holes slightly smaller than the nail diameter in the bottom side of each T. The holes should extend into the boom. The holes shouldn't be too small, since little resistance is needed to keep the pin in place. Take care

## Measurement Reveals Antenna Pattern Details

Doug Howard, VE6CID, measured Jerry Clement's take-down Yagi antenna in Trung Nguyen's Advanced Test Lab (**www.appliedtestlab.com**). The test chamber (Figure A) has ferrite absorber material on the walls, and is certified for Electromagnetic Compatibility (EMC) measurements. Figure B shows an azimuth pattern measured with 3 meters separation (this is a quasi-nearfield test) between the Yagi and the calibrated (red) test antenna. Jerry made his take-down Yagi as compact as possible, so he omitted the eight-turn coax common-mode choke/balun used in Joe Leggio's original design. The test is precise enough to reveal some asymmetry on the antenna pattern, likely because of the choke/balun omission.

**Figure A** — Cross-polarization test of the take-down Yagi antenna in the EMC measurement facility, viewed from the (red) calibrated test antenna.

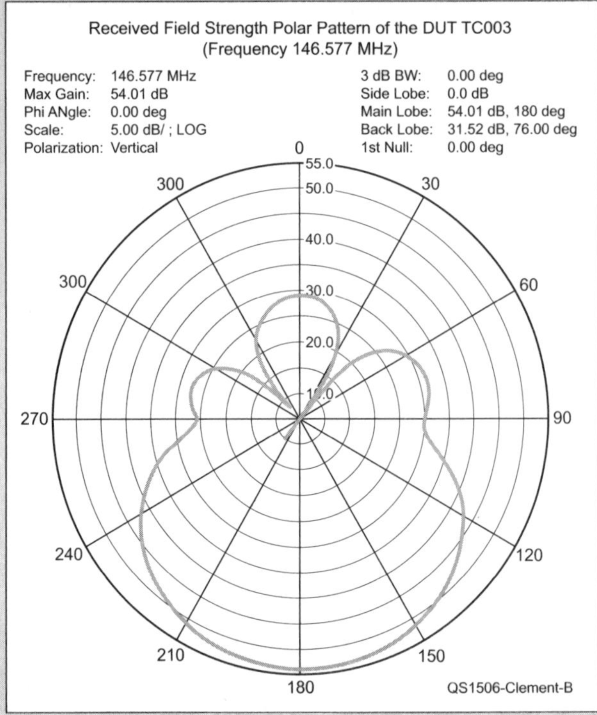


**Figure B** — Azimuth antenna pattern shows some asymmetry, likely because the common-mode choke was omitted.

not to allow the hole to protrude out the opposite side. Then lightly tap a pin into the drilled hole until the head is flush with the surface of the T. Install two pins into each coupler, one on either end of the coupler.

For the handle, cut a 4¼-inch long piece of the ½-inch Schedule 40 PVC pipe, and dress the ends square. Immediately behind the ½-inch boom protruding from the reflector assembly, wrap turns of electrical tape, one on top of the other, the width of the tape on the boom until the PVC handle just slides over tape wrap with little resistance. Place a second wrap of electrical tape the width of the tape at the rear of the boom. Space these two built-up wraps so that when the handle is positioned over them, there is a ¼ inch or so of space on either end. Apply hot glue from a hot glue gun into the cavity at one end of the handle until full. Let the glue cool before repeating this procedure at the other end of the handle. Dress the handle further, giving it a store-bought look, by cutting a piece of shrink-tube the required length to cover the handle. Shrink it into place over the handle with your heat gun.

**For safety, round off the ends of each element with your serrated scissors, then file until smooth.**

Solder the RG-58 coax and hairpin match in place (Figure 4). Strip back the coax the required amount and position the ends in the holes you drilled in the element blades. With the 5-inch length of solid #14 AWG copper wire, form a U that has a ¾-inch spacing. With a pair of sharp-nose pliers, form short 90 degree bends in the ends, and position the ends into the element holes along with the coax ends. Position the hairpin match parallel to the boom in front of the driven element. Temporarily slide the element keepers back a bit so they are not damaged by the heat of your soldering iron.

I used a 4-foot length of coax on my antenna. I also added a tie-wrap to secure the coax to the boom immediately behind the driven element T. Dress the area with shrink-tube to help keep the coax secure and to give the antenna a pleasing finished look. Once you've adjusted the hairpin match, place shrink-tube over it and the boom, securing the hairpin match in place. My antenna has an SWR of 1.1 to 1 at my "foxhunting" frequency of 146.565 MHz.

I built an "element winder" from a 2-inch-long piece of ¾-inch diameter Schedule 40 PVC pipe. Cut a slot the width of the element blade through the side at one end. Place the end of an element in the winder slot and wind the element on itself until you reach the stop. Secure the wound coils with 9⁄16-inch mini binder clips from Staples.[4]

**Notes**

[1] *The ARRL Antenna Book,* ARRL order no. 6948, available from your ARRL dealer, or from the ARRL Store, Telephone toll-free in the US 888-277-5289, or 860-594-0355, fax 860-594-0303; **www.arrl.org/shop/**; **pubsales@arrl.org**.

[2] **theleggios.net/wb2hol/projects/rdf/tape_bm.htm**

[3] **www.arrl.org/qst-in-depth**

[4] **www.staples.com/Staples-15348-Mini-Binder-Clip-12-PK-Black/product_779991**

Photos by the author.

International ARRL member Jerry Clement, VE6AB, has been a licensed Amateur Radio operator since 1992. Jerry is a machinist who owns and operates a machine shop where he builds scale models and makes mobile antennas. Jerry also specializes in automatic controls for refrigeration systems. Currently he provides technical support to agricultural clients located throughout Western Canada. An HF mobile enthusiast, he builds mobile antennas and evaluates their performance. He enjoys working through Amateur Radio satellites with antennas built in his shop. Jerry is also a photographer, and when not doing event work, enjoys photographing landscapes, wildlife, and nature scenes, as he explores the back roads of southern Alberta. You can reach Jerry at 3812 14 Ave NE, Calgary, AB T2A 7L6, Canada or **stormchaser@shaw.ca**.

**For updates to this article, see the *QST* Feedback page at www.arrl.org/feedback.**

# Build a Portable Groundplane Antenna

Need a better antenna for your hand-held radio? Here's the answer.

By Zack Lau, KH6CP/1
ARRL Laboratory Engineer

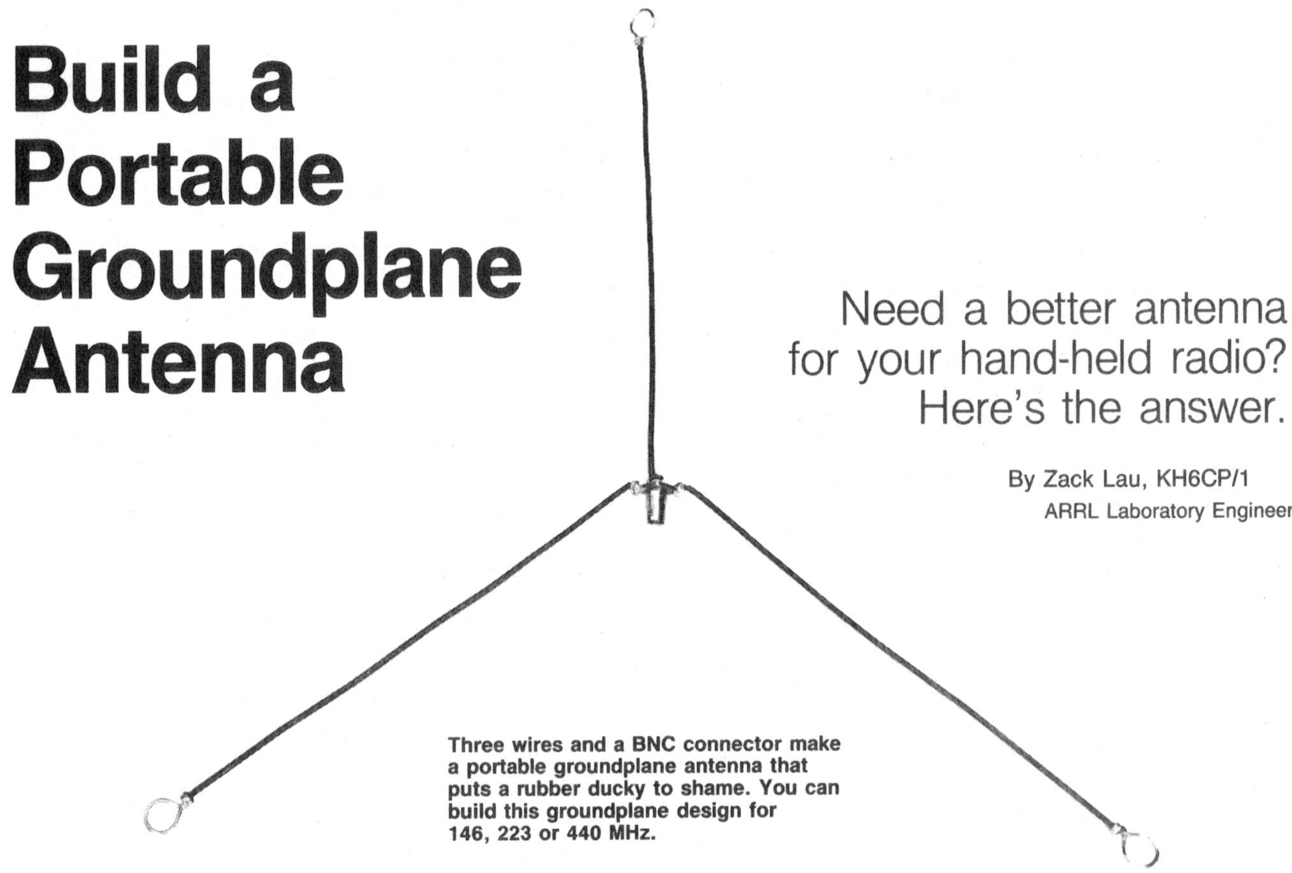

**Three wires and a BNC connector make a portable groundplane antenna that puts a rubber ducky to shame. You can build this groundplane design for 146, 223 or 440 MHz.**

The rubber ducky antennas common on hand-held VHF and UHF transceivers work fine in many situations. That's no surprise, considering that repeaters generally reside high and in the clear so you and your hand-held don't have to! Sometimes, though, you need a more efficient antenna that's just as portable as a hand-held. Here's one: A simple *groundplane* antenna you can build—for 146, 223 or 440 MHz—in no time flat. It features wire-end loops for safety (sharp, straight wires are hazardous) and convenience (its top loop lets you hang it off high objects for best performance).

**What You Need to Build One**

All you'll need are wire (single-conductor, no. 12 THHN), solder and a female coax jack for the connector series of your choice. Many hardware stores sell THHN wire—that is, thermal-insulation, solid-copper house wire—by the foot. Get 7 feet of wire for a 146-MHz antenna, 5 feet of wire for a 223-MHz antenna, or 3 feet of wire for a 440-MHz antenna.

The only tools you need are a 100-watt soldering iron or gun; a yardstick, long ruler or tape measure; a pair of wire cutters; a ½-inch-diameter form for bending the wire loops (a section of hardwood dowel or metal tubing works fine), and a file (for smoothing rough cut-wire edges and filing the coax jack for soldering). You may also find a sharp knife useful for removing the THHN's insulation.

**Building It**

To build a 146-MHz antenna, cut three 24-5/8-inch pieces from the wire you bought. To build a 223-MHz antenna, cut three 17-5/8-inch pieces. To build a 440-MHz antenna, cut three 10-5/8-inch pieces.

The photos show how to build the antenna, but they may not communicate why the cut lengths I prescribe are somewhat longer than the finished antenna's wires. Here's why: The extra wire allows you to bend and shape the loops by hand. The half-inch-diameter loop form helps you form the loops easily.

*Make the End Loops First*

Form an end loop on each wire as shown in Fig 1. Strip exactly 4 inches of insulation from the wire. Using your ½-inch-diameter form, bend the loop and close it—right up against the wire insulation—with a two-turn twist as shown in the bottommost example in Fig 1. Cut off the excess wire (about ½ inch). Solder the two-turn twist. Do this for each of the antenna's three wires.

*Attach the Vertical Wire to the Coax-Jack Center Pin*

Strip exactly 3 inches of insulation from the unlooped end of one of your wires and follow the steps shown in Fig 2. Solder the wire to the connector center conductor. (Soldering the wire to a coaxial jack's center pin takes considerable heat. A 700- to 750-°F iron with a large tip, used in a draft-free room, works best. Don't try to do the job with an iron that draws less than 100 watts.) Cut off the extra wire (about ½ inch).

*Attaching the Lower Wires to the Connector Flange*

Strip exactly 3 inches of insulation from the unlooped ends of the remaining two wires. Loop their stripped ends—right up to the insulation—through opposing mounting holes on the connector flange. Solder them to the connector. (You may need to file the connector flange to get it to take solder better.) Cut off the excess wire (about 2¼ inches per wire). This completes construction.

**Adjusting the Antenna for Best Performance**

Bend the antenna's two lower wires to form 120° angles with the vertical wire. (No, you don't need a protractor: Just position the wires so they just about trisect a circle.) If you have no means of measuring SWR at your antenna's operating frequency, stop adjustment here and start enjoying your antenna! Every hand-held I

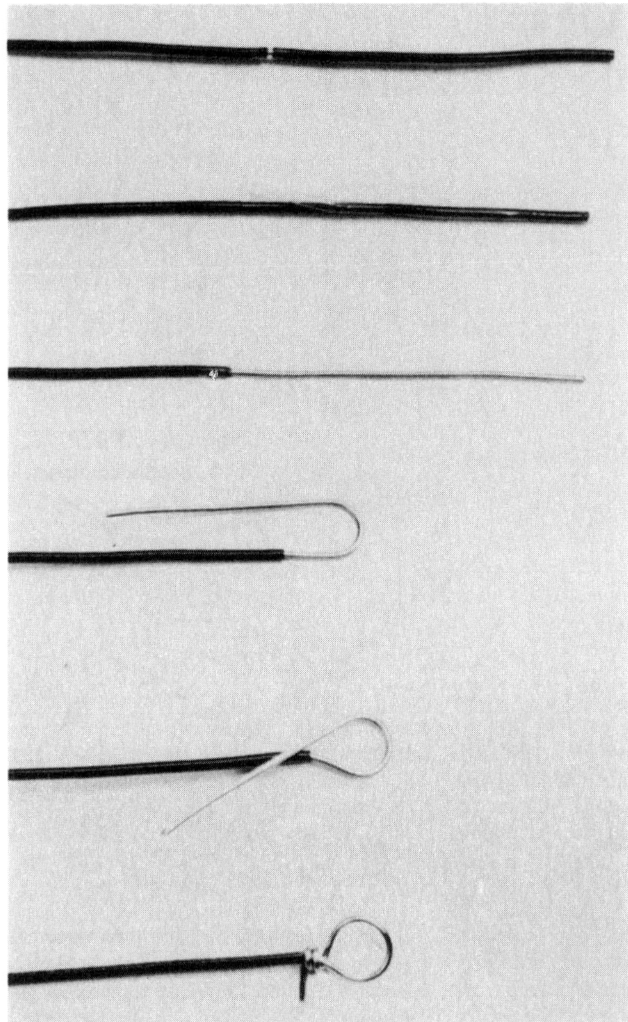

Fig 1—Making loops on the antenna wires requires that you remove exactly 4 inches of insulation from each. Stripping THHN insulation is easier if you remove its clear plastic jacket first.

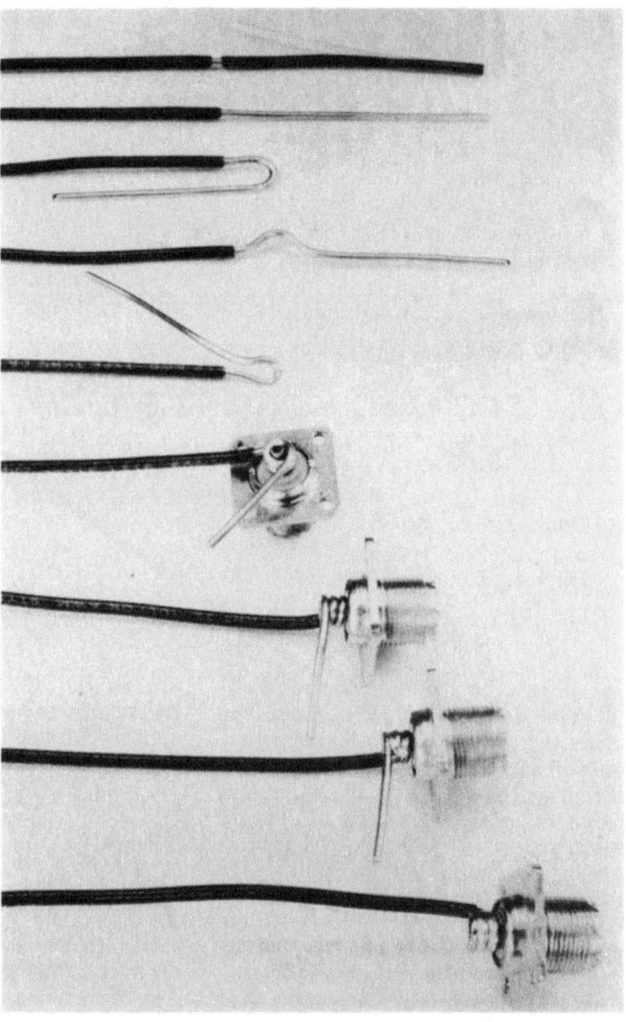

Fig 2—Remove exactly 3 inches of insulation to attach the vertical wire to the coax connector center pin. This photo shows an SO-239 (UHF-series) jack; the title photo shows a BNC jack. Use whatever your application requires.

**What's a Groundplane?**

This article emphasizes how to build and adjust a groundplane antenna for better communication at 146, 223 or 440 MHz. You can find out the technical details of *how* groundplane antennas work in Chapter 2 of *The ARRL Antenna Book*, available from your dealer and The ARRL Bookshelf.—*WJ1Z*

know of should produce ample RF output into the impedance represented by the antenna and its feed line.

Adjusting the antenna for minimum SWR is worth doing if you have an SWR meter or reflected-power indicator that works at your frequency of interest. Connect the meter in line between your handheld and the antenna. Between short, identified test transmissions—on a simplex frequency—to check the SWR, adjust the angle between the lower wires and the vertical wire for minimum SWR (or reflected power). (You can also adjust the antenna by changing the length of its wires, but you shouldn't have to do this to obtain an acceptable SWR.) Before considering the job done, test the antenna in the clear to be sure your adjustments still play. (Nearby objects can detune an antenna.)

**Plug and Play**

As you use the groundplane, keep in mind that its coax connector's center pin wasn't made to bear weight and may break if stressed too much. Barring that, your groundplane should require no maintenance at all.

There you go: You may not have built a monument to radio science, but you've home-constructed a portable antenna that'll get much more mileage from your hand-held than its stock rubber ducky. Who said useful ham gear has to be hard or expensive to build?

# A Traveler's 2-Meter GP Antenna

Try this low-cost, knock-down portable ground-plane antenna for 2 meters. It is compact, and offers better performance than a "rubber duckie" can provide.

By Doug DeMaw, W1FB
ARRL Contributing Editor
PO Box 250, Luther, MI 49656

I'm sure you've been as frustrated as I have during trips away from home with a 2-meter hand-held transceiver, but without an antenna that would permit access to those elusive, slightly distant repeaters. A frequent cause of repeater inaccessibility is inefficient antenna performance, such as we may expect from the rubber duckie that has become a standard fixture for hand-helds. Unfortunately, the ground system for such a radiator is only the electronics inside the transceiver case. Also, the rubber duckie is a physically short ¼-wavelength antenna that is helically wound. Although this short antenna is entirely practical and desirable for most hand-held use, it has serious drawbacks. This is especially true when the operator (using the short antenna and low power—generally 1 to 5 watts) is located too far from a repeater.

What might we do to improve the range of our hand-held transceivers? The first thought that comes to mind may suggest the use of a portable 3- or 5-element Yagi antenna with a mast. This would be great, provided we were camping or visiting at a friend's residence. But what about motels, hotels and other restrictive environments? A Yagi antenna would be rather impractical, and it would be cumbersome to carry with us.

## Some Solutions

A number of alternatives to the rubber duckie exist. For example, we might construct a simple half-wavelength dipole antenna and suspend it vertically at the portable site (motel room, for example). Another approach would be to use a J-pole antenna, such as the fine portable unit described in *QST* by Aurick.[1] For those who would like to build a portable Yagi for hand-held use, there are the units described some years ago by Campbell of the ARRL staff.[2] But, what about a ground-plane type of antenna. . .something that can be dismantled after use and packed into a small container? Such an antenna can be built with a base unit to permit the system to be placed on a dresser or table, near a window, in a hotel or motel room (remember that many buildings have steel frameworks, and there is disruptive plumbing and electrical wiring in the walls). This article describes a practical, compact 2-meter ground-plane antenna. It may be used also on 220 and 432 MHz with minor modifications.

## Ground Plane Details

Fig 1 is a pictorial view of the antenna I developed for my use with a 2-meter hand-held. The drawing shows the assembled unit. A wooden base is used to support the upper part of the antenna. A 6-inch section of 7/8-inch-OD PVC pipe (¾-inch ID) serves as the vertical support column for the radials and radiator. A 7/8-inch-ID copper plumbing cap fits over the PVC tubing to serve as a mounting fixture for the antenna elements. An impedance-matching loop connects between one radial and the bottom of the radiator element (more on this later). A short length of 50-ohm coaxial cable exits from the base of the antenna for connection to the hand-held.

The antenna elements unscrew from the copper cap, the impedance-matching loop is taken off, and the PVC column pulls out of the wooden base. This permits the user to pack the antenna into a small bag or box for storage until it is used another time. The antenna elements are made from telescoping rods that are available as replacement parts for portable radios and TV sets. My rods are 4¾ inches long when collapsed. They extend to 20 inches. Longer rods are also available as surplus.

## Construction Data

Ordinary tools may be used for making this antenna. A break-down view of the supporting structure is shown in Fig 2. Perhaps the most difficult part of the job is drilling the side holes in the copper cap (item A of Fig 2). The no. 4 holes (four each) should be 90 degrees apart for best appearance, but it's not a critical spacing for good performance. Lay out the hole positions by marking the spots with a laundry pen. Center punch the cap where each hole will be drilled. This will make it easier to drill into the curved surface of the cap. The top-center hole will be easy to drill. It should be larger than the four radial-element holes to accommodate a no. 4 insulating washer with a shoulder. The radiator element must be insulated from the cap, but the four radial elements are common to the cap. A flat fiber washer is needed over the top hole inside the cap to mate with the shoulder washer that is installed from the top of the cap.

If you can't locate a shoulder and flat insulating-washer set, you may drill a ½-inch hole in the top of the cap, then glue a piece of unclad PC board or plastic inside the cap to cover the ½-inch hole. This alternative insulator may then be drilled for a no. 4 screw.

The antenna rods I purchased from Mouser Electronics are threaded at the base ends for a no. 3 metric screw. I did not have the proper screws for the rods, so I rethreaded the rods with a no. 4-40 tap.

[1]Notes appear on page 95.

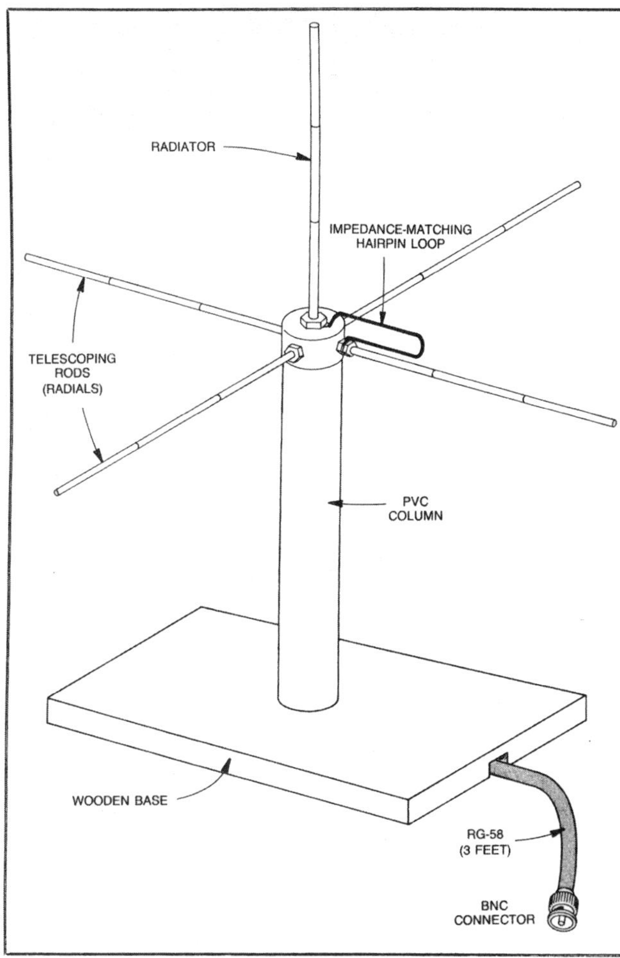


Fig 1—Pictorial drawing of the assembled portable 2-meter ground-plane antenna. The elements may be collapsed and removed from the vertical support member for storage. The impedance-matching loop at the top may also be removed easily, and the vertical column can be pulled from the wooden base. These features permit packing the antenna in a small container when carrying it afield.

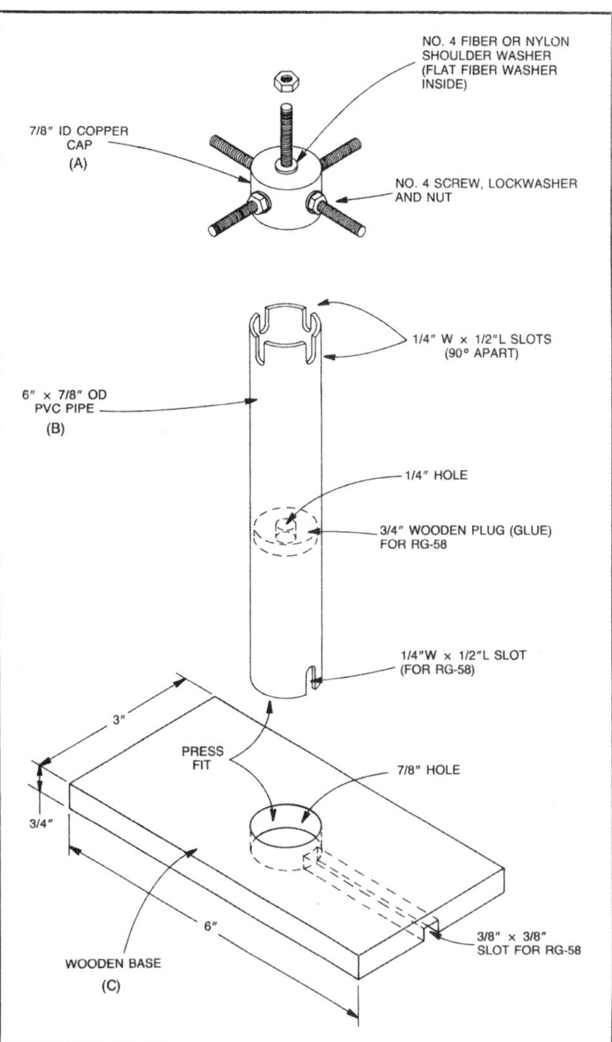


Fig 2—Break-down view of the ground-plane antenna. Ordinary materials are used to construct this unit. All of the components are easy to obtain.

This enabled me to use five 4-40 screws, 5/8 inch long, for attaching the five antenna rods. A solder lug is used inside the cap, for the radiator, and one solder lug is attached inside the cap by means of one of the radial-element screws. The solder lugs provide connection points within the cap for the RG-58 coaxial cable: The shield braid is common to the radials and the center conductor is attached to the bottom of the vertical-radiator element.

You will notice that the PVC column (part B of Fig 2) has four slots at the top end. These are necessary because the cap will not fit over the end of the PVC column without them, owing to the presence of the four radial-element screw heads. Therefore, the slots should also be 90 degrees apart, or at least aligned with the four side screws on the cap. I cut the four slots with a router and small bit. A hack saw may be used for slotting the column. The pieces to be removed from the slots can be snapped off. A round file will be useful in rounding off the bottoms of the slots.

I glued a wooden plug inside the PVC column about midway through the pipe (dashed lines in Fig 2). This plug has a ¼-inch center hole through which the RG-58 cable is passed. The plug functions as a strain-relief device to minimize twisting of the cable within the cap section. A fifth slot is located at the bottom of the PVC column. It allows the RG-58 to lie in the rectangular slot in the wooden base (part C of Fig 2) of the antenna. Without the PVC slot, and that in the base C, the cable would prevent the base from resting flat on the table.

I cut the slot on the underside of the wooden base by means of my router. This job can be accomplished also by using a wood gouge or chisel. The sides of the slot can be cut with a hacksaw blade to ease the chore. The center hole in the base should be small enough to allow a snug fit between the PVC column and the base.

Part of this effort can be deleted if you wish to install a BNC female jack on the side of the PVC column. A model I developed in 1984 was structured in that manner. This would eliminate having the cable pass through the wooden base and along the rectangular slot. A connecting cable could then be used between the antenna and the hand-held, thereby eliminating the permanently attached RG-58 line.

The wooden base does not have sufficient weight to keep the assembled antenna upright when there is stress on the coaxial cable. My base has three 5/16- × 2¼-inch holes drilled laterally through the nongrooved end of the base. Two similar channel holes are located on the grooved end of the base plate. The five holes are filled with lead to increase the weight of the antenna base. I used a propane torch to melt some large fishing

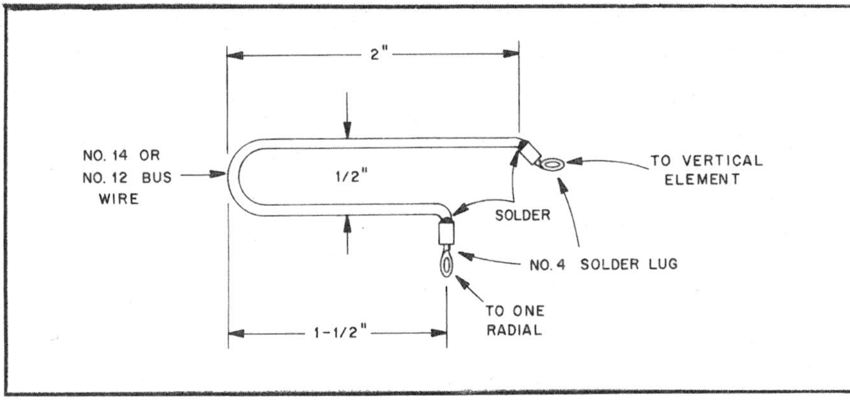


Fig 3—Details of the matching device for the antenna. This impedance-matching loop transforms the 30-ohm feed impedance to 50 ohms for use with RG-58 cable. The loop connects between the radials and the bottom of the vertical radiator.

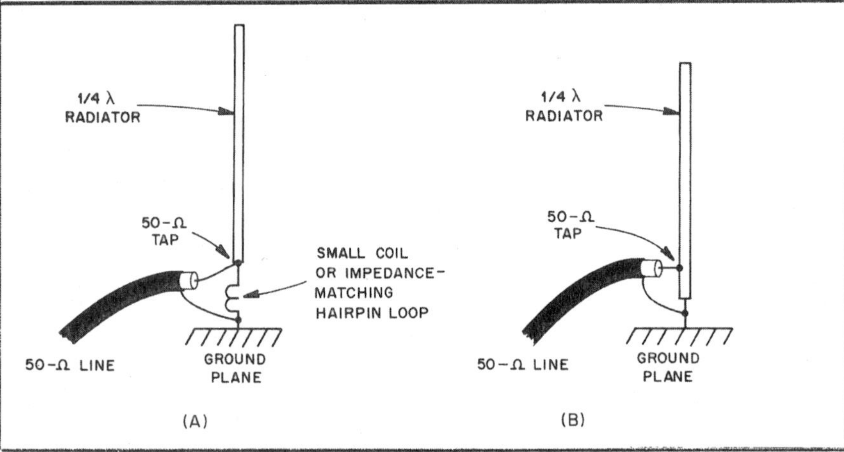


Fig 4—Representation of how the impedance-matching loop provides a 50-ohm tap point on the antenna (A). The equivalent circuit at B illustrates the tap point on a quarter-wavelength radiator (see text).

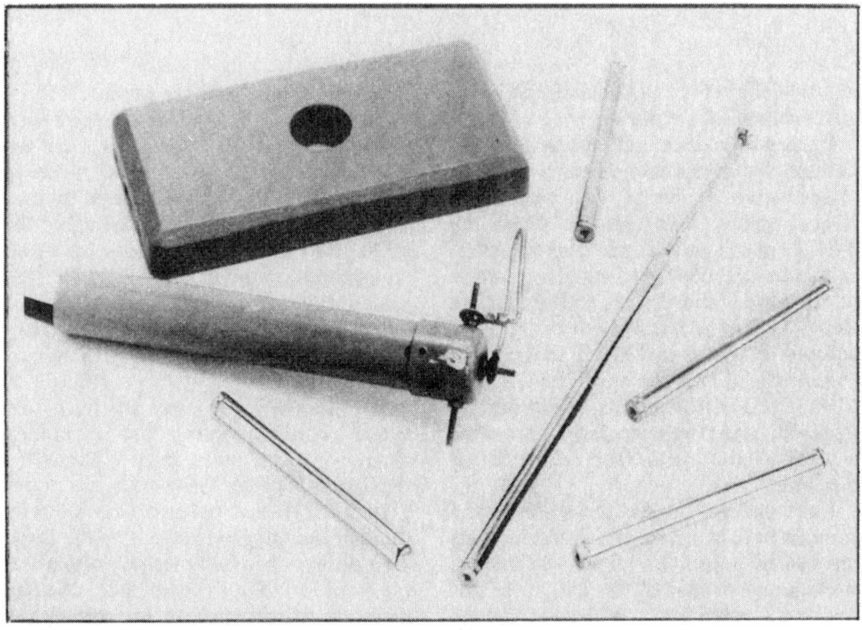

Fig 5—Photographic view of the disassembled antenna. The parts may be stored in a small box or bag for convenient transportation.

sinkers in a small tin can. The can was bent at one point to form a pouring lip. The lead was poured into the antenna-base holes. This corrected the "tipsy" problem. Alternatively, you may choose to hold the antenna base in place on a table or desk with a C clamp.

### Electrical Considerations

Be cautious about the method for connecting the RG-58 cable inside the copper cap. The shield braid must not come in contact with the center conductor of the line. I keep the exposed part of the braid short, while making the center-conductor piece (with its insulation) about ½ inch long. The solder lug for the radials points downward, and the braid is soldered to the lower end of the lug. Just before the cap is installed on the PVC column I squeeze a generous amount of Dow Corning silicone cement or GE Silastic® compound into the cap. This helps to prevent the shield braid from shorting to the cable center conductor during stress. It also affixes the cap to the PVC column. Allow at least 24 hours for the sealant to set before using the antenna.

### Impedance-Matching Loop

The feed impedance of a quarter-wavelength ground-plane antenna is on the order of 30 ohms. For the purpose of matching the antenna to a 50-ohm line, it is necessary to employ some type of matching device. The U-shaped hairpin loop seen in Fig 3 elevates the impedance from 30 to 50 ohms for matching the system to the RG-58 line. One end of the loop is attached outside the cap, under one radial. The remaining end of the loop is connected outside the cap, to the bottom of the vertical radiator. A solder lug is used at each end of the loop for convenience of attachment. In effect, the loop provides the electrical equivalent of a feed line being tapped up on the low-impedance end of the vertical element. Therefore, the vertical element is common to the radials when using the loop. Fig 4 shows the relationship between the two methods described.

Smaller loops or short straps may be made for use on 220 and 432 MHz. I have not worked out the dimensions, but with the aid of an SWR indicator it should be an easy matter to develop the dimensions experimentally. Fortunately, the antenna rods I purchased collapse sufficiently for use on the two higher amateur bands. You may want to keep this in mind when you obtain your rods. If you are a 6-meter enthusiast, there is no reason why this design cannot be used for that band by providing longer rods and a larger loop. I suggest that a heavy metal base be used for 6-meter antennas. This will keep the antenna from tipping over when the longer elements are extended.

### Adjustment and Use

I extend the radials of my antenna to

20 inches. Next, I use an SWR bridge to set the length of the vertical radiator for an SWR of 1. A small metal file may be used to nick the radiator rod so that you can set it for the correct length later on. I found that I had to lengthen my vertical element by approximately 2 inches to provide an adequate range of adjustment. This I did by force-fitting and crimping the tip of another rod to the existing vertical radiator.

How does this antenna compare to a rubber duckie? Well, my report is based on relative comparisons, owing to the unknown character of the S-meter calibration of my hand-held. While using the transceiver inside my radio room at Luther, I brought up the 146.79-MHz repeater at Manistee, Michigan (K8CEB), which is some 30 airline miles from this QTH. The repeater signal registered ⅓ scale when using the rubber duckie. I then placed the ground-plane antenna in the same spot and attached it to the hand-held. The repeater signal registered full scale, and it was full quieting. A check of the WD8RZL repeater, about 15 airline miles away, showed less-dramatic results. With the rubber duckie it registered ¾ scale on the S meter, and went to full scale when using the ground-plane antenna. Although I didn't measure the difference in performance in decibels, in any event, the ground-plane antenna is superior to the rubber duckie in overall performance. It is well worth the effort it took to construct it.

### In Conclusion

It may seem like this was a lot of ado about a simple, established antenna concept. However, from the viewpoint of portability (Fig 5) and improved operation afield, the story is worth telling. I am sure that many of you can think of better ways to fabricate an antenna of this kind. I chose the design shown here because all of the materials are available locally at hardware and lumber stores. The antenna rods are offered by many surplus mail-order dealers. A variety of mounting styles are available, such as rods that have flat ends with a screw hole. Others have a small threaded stud on one end. This opens the door to a host of construction techniques for this antenna. However your end product looks, I'm sure you will be glad you constructed a portable GP antenna!

#### Notes

[1]L. Aurick, "The Timeless J Antenna," *QST*, Nov 1982, pp 40-41.

[2]E. L. Campbell, "Two Toter," *QST*, Jul 1971, pp 23-25, and "Portable Beam for Two Meters," *QST*, Oct 1974, pp 36-37.

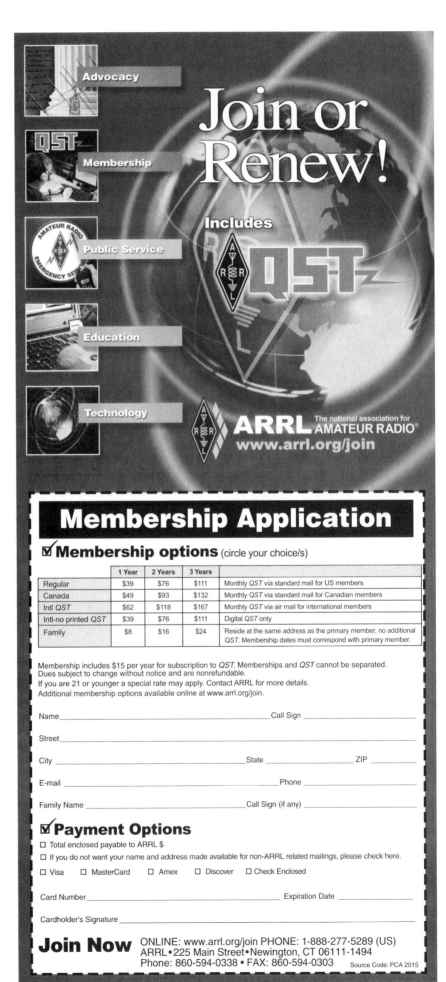

# A Portable Quad for 2 Meters†

Backpacking, boating or mountaintopping? Invest an afternoon's work and pack this novel directional gain antenna on your next expedition.

By R. J. Decesari,* WA9GDZ/6

Last year, while I was "hilltopping" in the San Diego area with my 2-meter fm transceiver, a band opening occurred in which stations from Los Angeles, Santa Barbara and further points north were copied on simplex frequencies. Establishing solid communications with the built-in quarter-wave whip antenna and 1-watt power of the transceiver (with weakening batteries) was rather difficult, even with the opening. Because of my intense desire to communicate with these DX stations, a need for either a directional gain antenna or a power amplifier was established. Since I didn't particularly desire toting and charging additional batteries for an amplifier, I set this concept aside. I then took a closer look at improving the antenna. This novel portable antenna configuration evolved from many hours of thinking and tinkering in my workshop.

Initial efforts to design a collapsible antenna centered on a conventional four-element Yagi configuration. Several models of the Yagi, whose elements all opened simultaneously, proved to be a nightmare in bell cranks and lever arms. From this attempt, I decided that all the elements should still be attached to a main boom, but the operator would open the elements individually during antenna set-up, thus eliminating the push rods and cranks. The Yagi design, with the elements folding on top of each other to minimize space, was still rather large considering element spacing and other required mechanical appendages and dimensions. At about this time, I happened to spot a big 20-meter quad while driving to work and immediately started to ponder the possibilities of using a quad for the intended portable antenna.

With only two elements, the quad

*3941 Mt. Brundage, San Diego, CA 92111

†A patent is pending on the antenna system described in this article; commercial application of this construction technique is prohibited.

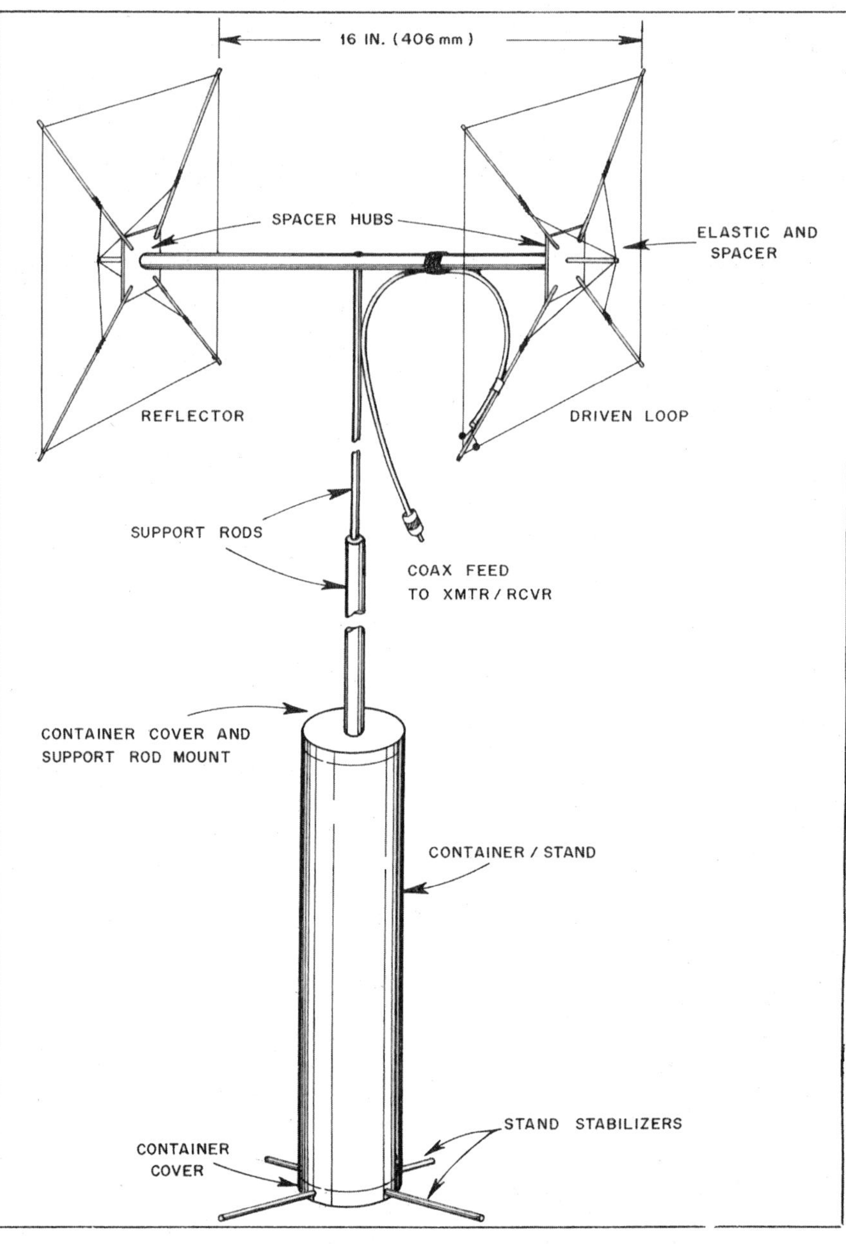


Fig. 1 — The basic portable quad assembly. The author used an element spacing of 16 in. (406 mm) so that the quad spacers would fold neatly between the hubs.

provides an excellent front-to-back ratio, as well as about 6 dB of forward gain. With a two-element quad, the element spacing for optimum reflector performance is between 0.15 λ and 0.2 λ. That works out to about 12 and 16 inches (305 and 406 mm) at 2 meters. Not a bad overall size for a 2-meter antenna! Now the problem was how to support the square loops. A quick lesson in geometry revealed that if an "X" configuration of spacers were used to support 144-MHz loops, then each leg of the "X" would also be about 16 inches! All that was left to do was design a center hub that would allow the spacers to fold to the longitudinal axis of the boom and the basic problem would be solved. Consequently, the garage workshop was put into overtime service and the preliminary model of the brainchild was fabricated.

Fig. 2 — The portable quad in stow configuration. Two long dowels are used as support rods. Four smaller dowels are used to stabilize the container.

## A Quad is Born

Figs. 1 and 2 show the basic portable quad. Both driven and reflector elements fold back on top of each other, resulting in a structure about 17 inches (432 mm) long: The wire loop elements may be held in place around the boom with an elastic band. To support the antenna once it has been erected, the container is used as a stand. To provide more stability, four small removable struts slip into holes in the base of the container. Both the support rods and struts fit inside the container when the antenna is disassembled.

I have used two different methods of keeping the quad spacers erect. Both methods are successful. Fig. 3 shows the quad spacers held open by spring-steel clips. Each clip is fabricated from an ordinary paper binder with a hole drilled in it to allow it to be attached to the quad spacer. The clip is compressed and slid down the quad spacer until it engages the hub. This provides a rigid mechanical support to hold the spacer open when in use as well as allowing it to pivot back for storage in the container. Fig. 4 shows a slightly different method: A mechanical stop is machined into the hub, and elastic bands are used to hold the spacers erect. The bands are attached to an additional strut to hold the spacers open. When not in use, the strut pulls out and sits across the hub, and the spacers can be folded back. Details of each method are shown in Fig. 5.

The clip and hub assembly is possibly easier for the home builder to fabricate, with the exception of drilling the hole in the spring steel. A high-speed-steel or carbide-tipped drill set is required, since the spring steel is an extremely tough and brittle material. Care must be taken when drilling the holes since the clip material will tend to crack. It is recommended that the builder start with a small-diameter drill and proceed to sequentially larger drill diameters until the final diameter is reached. The clip should be expanded and

Fig. 3 — Paper-binder spring clips are used in this version of the quad to hold the spacers erect.

Fig. 4 — This version of the portable quad uses mechanical stops machined into the hub; elastic bands hold the spacers open.

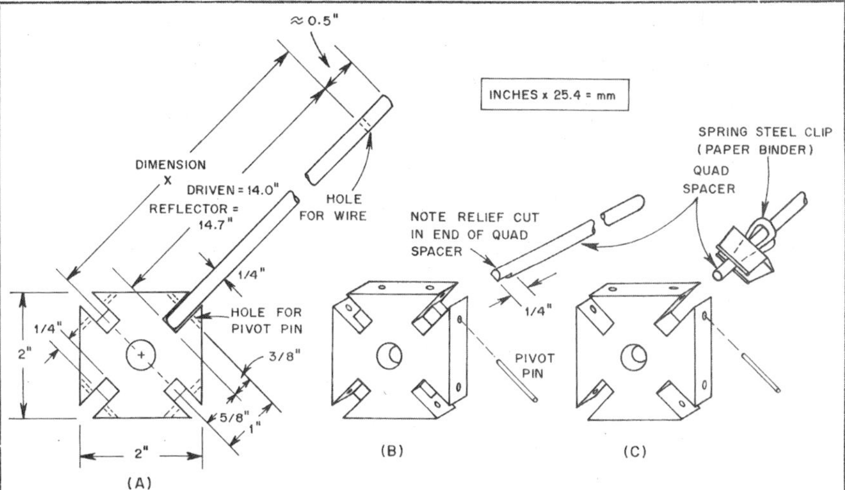


Fig. 5 — A detail of the spacer hub with spacer lengths for the director and reflector is shown at A. The hub is made from 1/4-inch (6.4-mm) plastic or hardwood material. The center-hole diameter can be whatever is necessary to match the diameter of your boom. The version of the hub with mechanical stops and elastic bands is shown at B. At C is the spacer hub version using spring clips.

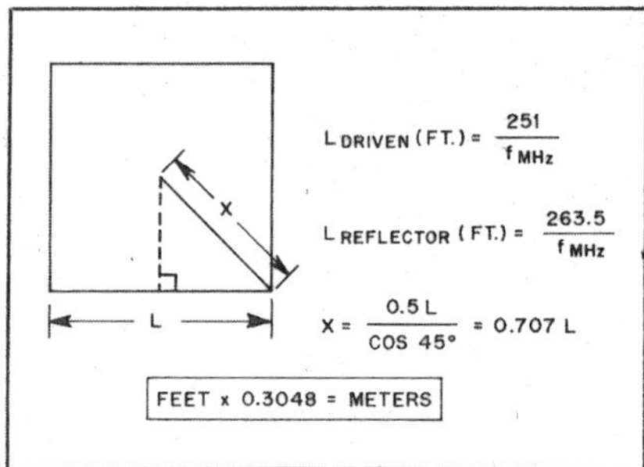


Fig. 6 — Quad loop dimensions. Dimension X is the distance from the center of the hub to the hole drilled in each spacer for the loop wire. At 146 MHz, dimension X for the driven element is 1.216 feet (14.6 inches), and dimension X for the reflector is 1.276 feet (15.3 inches).

fitted over a 1/4-inch (6.4-mm) piece of wood to be used as a drilling back. Use of a light oil is recommended to keep the drill tip cool.

**Building Materials**

The portable quad antenna may be fabricated from any one of several plastic or wood materials. The most inexpensive method is to use wood doweling, available at most hardware stores. Wood is inexpensive and easily worked with hand tools; 1/4-inch (6.4-mm) doweling may be used for the quad spacers, and 3/8- or 1/2-inch (9.5- or 12.7-mm) doweling may be used for the boom and support elements. A hardwood is recommended for the hub assembly, since a softwood may tend to crack along its grain if the hub is impacted or dropped. Plastics will also work well, but the cost will rise sharply if the material is purchased from a supplier. Plexiglas is an excellent candidate for the hub. Using a router and hand tools, I manufactured a set of Plexiglas hubs with no difficulty. Fiberglass or phenolic rods are also excellent for the quad elements and support.

The loops were made with no. 18 AWG insulated stranded copper wire, although enameled wire may also be used. If no insulation is used on the wire and wood doweling is used for the spacers, a coat of spar varnish in and around the spacer hole through which the wire runs is recommended. The loop wire terminates at one element by attaching to heavy-gauge copper-wire posts inserted into tightly fitting holes in the element. For the driven element, two posts are used to allow the RG-58/U feed-line braid and center conductor to be attached. A single post is used on the reflector to complete the loop circuitry.

The first model of this antenna had a tuning stub attached to the reflector loop. This allowed a certain degree of reflector tuning to maximize its performance. However, I discovered a computer maximization of quad loop and spacing dimensions.[1] This data was used in my subsequent 2-meter quad designs, and has simplified the antenna by eliminating the need for a reflector tuning stub. Fig. 6 shows quad dimensions derived from this data. The quads described in this article have been designed for 146 MHz, but the basic loop size equations will allow the builder to construct a model to any desired frequency in the 2-meter band to maximize results.

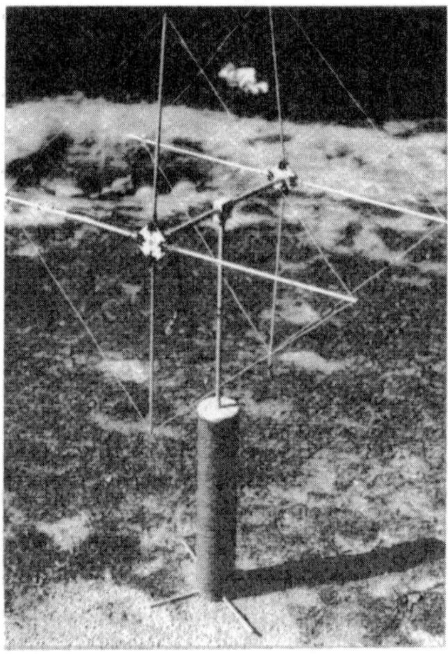

Fig. 7 — A vertically polarized portable quad. The feed point is at the extreme left of the photograph.

The author with the fully erected portable quad antenna. The bottom stand is also used as a storage container.

The storage container was made from a heavy cardboard tube originally used to store roll paper. Any rigid cylindrical housing of the proper dimensions may be used. Two wood end pieces were fabricated to cap the cardboard cylinder. The bottom end piece is cemented in place and has four holes drilled at 90° angles around the circumference. These holes hold 4-inch (102-mm) struts, which provide additional support when the antenna is erected. The top end piece is snug fitting and removable. It is of sufficient thickness (about 5/8 inch or 16 mm) to provide sufficient support for the antenna-supporting elements. A mounting hole for the supporting elements is drilled in the center of the top end piece. This hole is drilled only about three-quarters of the way through the end piece and should provide a snug fit for the antenna support. One or more antenna support elements may be used, depending on the height the builder wishes to have. Keep in mind, however, that the structure will be more prone to blow over, the higher above the ground it gets! Doweling and snug-fitting holes are used to mate the support elements and the antenna boom.

**Polarization and Performance**

The antennas shown in Figs. 1 through 4 all have 45° diagonal polarization. This is a compromise between vertical and horizontal polarization that allows both fm and ssb/cw (which is usually horizontally polarized) to be worked on 2 meters. Fig. 7 shows another version of the antenna, built for vertical polarization. Although analytical antenna-pattern and gain tests have not been conducted, the portable quad displays an excellent front-to-back ratio as well as gain. The antenna has been used in the field with very satisfying results. The best example of the performance of the antenna was demonstrated by comparison to a 5/8-wave whip antenna. In this demonstration, the 5/8-wave whip was placed on a table top inside the ham shack and excited with 15 watts. From a location in San Diego, the 5/8-wave whip was unable to trigger any of the Los Angeles repeaters about 150 miles to the north. With the portable quad sitting on the same table, full-quieting access was gained to the Los Angeles repeaters.

This antenna design provides a compact package for a directional-gain antenna ideally suited for portable operation. Furthermore, it can be built from readily available and inexpensive materials. I would like to thank my father-in-law for his encouragement and my wife Sue for her patience and indulgence.

[1] "Optimum gain element spacing found for the quad antenna," *World Radio News*, March 1978.

# The DBJ-2: A Portable VHF-UHF Roll-Up J-pole Antenna for Public Service

*WB6IQN reviews the theory of the dual band 2 meter / 70 cm J-pole antenna and then makes detailed measurements of a practical, easy to replicate, "roll-up" portable antenna.*

**Edison Fong, WB6IQN**

It has now been more than three years since my article on the dual band J-pole (DBJ-1) appeared in the February 2003 issue of *QST*.[1] I have had over 500 inquires regarding that antenna. Users have reported good results, and a few individuals even built the antenna and confirmed the reported measurements. Several major cities are using this antenna for their schools, churches and emergency operations center. When asked why they choose the DBJ-1, the most common answer was value. When budgets are tight and you want a good performance-to-price ratio, the DBJ-1 (*D*ual *B*and *J*-pole–*1*) is an excellent choice.

In quantity, the materials cost about $5 per antenna and what you get is a VHF/UHF base station antenna with $\lambda/2$ vertical performance on both VHF and UHF bands. If a small city builds a dozen of these antennas for schools, public buildings, etc it would cost about $60. Not for one, but the entire dozen!

Since it is constructed using PVC pipe, it is UV protected and it is waterproof. To date I have personally constructed over 400 of these antennas for various groups and individuals and have had excellent results. One has withstood harsh winter conditions in the mountains of McCall, Idaho for four years.

The most common request from users is for a portable "roll-up" version of this antenna for backpacking or emergency use. To address this request, I will describe how the principles of the DBJ-1 can be extended to a portable roll-up antenna. Since it is the second version of this antenna, I call it the DBJ-2.

## Principles of the DBJ-1

The earlier DBJ-1 is based on the J-pole,[2] shown in Figure 1. Unlike the popular ground plane antenna, it doesn't need ground radials. The DBJ-1 is easy to construct using inexpensive materials from your local hardware store. For its simplicity and small size, the DBJ-1 offers excellent performance and consistently outperforms a ground plane antenna.

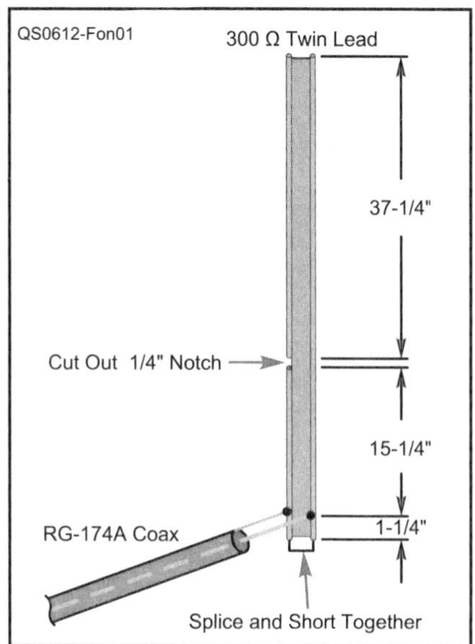


**Figure 1 — The original 2 meter ribbon J-pole antenna.**

Its radiation pattern is close to that of an ideal vertical dipole because it is end-fed, with virtually no distortion of the radiation pattern due to the feed line. A vertically polarized, center-fed dipole will always have some distortion of its pattern because the feed line comes out at its center, even when a balun is used. A vertically polarized, center-fed antenna is also physically more difficult to construct because of that feed line coming out horizontally from the center.

The basic J-pole antenna is a half-wave vertical configuration. Unlike a vertical dipole, which because of its center feed is usually mounted alongside a tower or some kind of metal supporting structure, the radiation pattern of an end-fed J-pole mounted at the top of a tower is not distorted.

The J-pole works by matching a low impedance (50 Ω) feed line to the high impedance at the end of a $\lambda/2$ vertical dipole. This is accomplished with a $\lambda/4$ matching stub shorted at one end and open at the other. The impedance repeats every $\lambda/2$, or every 360° around the Smith Chart. Between the shorted end and the high impedance end of the $\lambda/4$ shorted stub, there is a point that is close to 50 Ω and this is where the 50 Ω coax is connected.

By experimenting, this point is found to be about 1¼ inches from the shorted end on 2 meters. This makes intuitive sense since 50 Ω is closer to a short than to an open circuit. Although the Smith Chart shows that this point is slightly inductive, it is still an excellent match to 50 Ω coax. At resonance the SWR is below 1.2:1. Figure 1 shows the dimensions for a 2-meter J-pole. The 15¼ inch $\lambda/4$ section serves as the quarter wave matching transformer.

A commonly asked question is, "Why 15¼ inches?" Isn't a $\lambda/4$ at 2 meters about 18½ inches? Yes, but twinlead has a reduced velocity factor (about 0.8) compared to air and must thus be shortened by about 20%.

A conventional J-pole configuration works well because there is decoupling of the feed line from the $\lambda/2$ radiator element since the feed line is in line with the radiating $\lambda/2$ element. Thus, pattern distortion is minimized. But this only describes a single band VHF J-pole. How do we make this into a dual band J-pole?

## Adding a Second Band to the J-pole

To incorporate UHF coverage into a VHF J-pole requires some explanation. (A more detailed explanation is given in my February 2003 *QST* article.) First, a 2 meter antenna does resonate at UHF. The key word here is

[1]Notes appear on page 101.

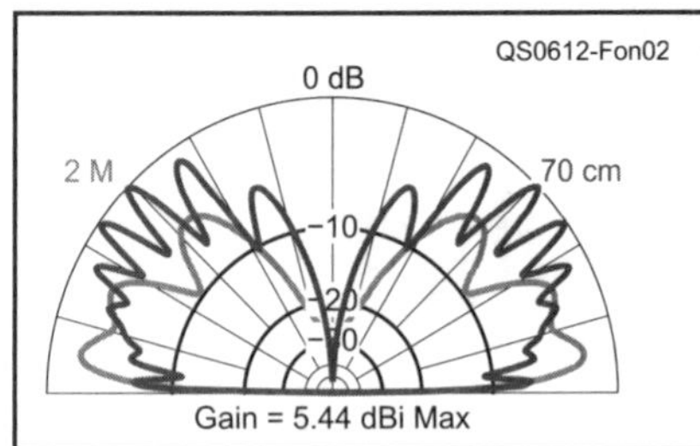


**Figure 2 — Elevation plane pattern comparing 2 meter J-pole on fundamental and on third harmonic frequency (70 cm), with the antenna mounted 8 feet above ground. Most of the energy at the third harmonic is launched at 44°.**

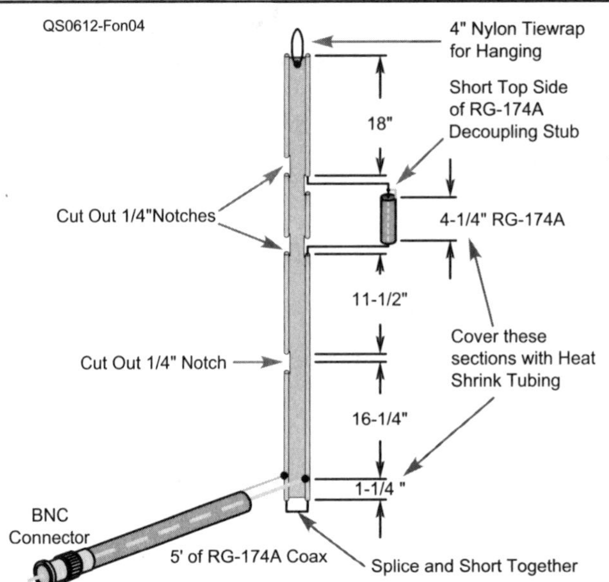


**Figure 4 — The dual-band J-pole modified for portable operation — thus becoming the DBJ-2. Note that the dimensions are slightly longer than those in Figure 3 because it is not enclosed in a PVC dielectric tube. Please remember that the exact dimensions vary with the manufacturer of the 300 Ω line, especially the exact tap point where the RG-174A feed coax for the radio is connected.**

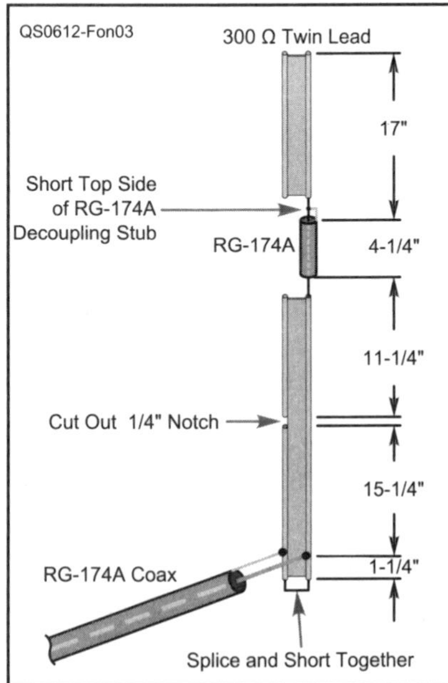


**Figure 3 — The original DBJ-1 dual-band J-pole. The dimensions given assume that the antenna is inserted into a ¾ inch Class 200 PVC pipe.**

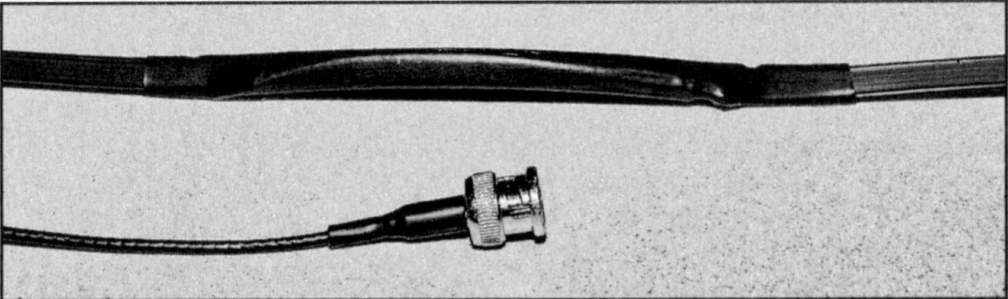

**Figure 5 — The λ/4 UHF decoupling stub made of RG-174A, covered with heat shrink tubing. This is shown next to the BNC connector that goes to the transceiver.**

*resonate*. For example, any LC circuit can be resonant, but that does not imply that it works well as an antenna. Resonating is one thing; working well as an antenna is another. You should understand that a λ/4 146 MHz matching stub works as a 3λ/4 matching stub at 450 MHz, except for the small amount of extra transmission line losses of the extra λ/2 at UHF. The UHF signal is simply taking one more revolution around the Smith Chart.

The uniqueness of the DBJ-1 concept is that it not only resonates on both bands but also actually performs as a λ/2 radiator on both bands. An interesting fact to note is that almost all antennas will resonate at their third harmonic (it will resonate on any odd harmonic 3, 5, 7, etc). This is why a 40 meter dipole can be used on 15 meters. The difference is that the performance at the third harmonic is poor when the antenna is used in a vertical configuration, as in the J pole shown in Figure 1. This can be best explained by a 19 inch 2 meter vertical over an ideal ground plane. At 2 meters, it is a λ/4 length vertical (approximately 18 inches). At UHF (450 MHz) it is a 3λ/4 vertical. Unfortunately, the additional λ/2 at UHF is out of phase with the bottom λ/4. This means cancellation occurs in the radiation pattern and the majority of the energy is launched at a takeoff angle of 45°. This results in about a 4 to 6 dB loss in the horizontal plane compared to a conventional λ/4 vertical placed over a ground plane. A horizontal radiation pattern obtained from *EZNEC* is shown in Figure 2. Notice that the 3λ/4 radiator has most of its energy at 45°.

Thus, although an antenna can be made to work at its third harmonic, its performance is poor. What we need is a simple, reliable method to decouple the remaining λ/2 at UHF of a 2 meter radiator, but have it remain electrically unaffected at VHF. We want independent λ/2 radiators at both VHF and UHF frequencies. The original DBJ-1 used a combination of coaxial stubs and 300 Ω twinlead cable, as shown in Figure 3.

Refer to Figure 3, and start from the left hand bottom. Proceed vertically to the RG-174A lead in cable. To connect to the antenna, about 5 feet of RG-174A was used with a BNC connector on the other end. The λ/4 VHF impedance transformer is made from 300 Ω twin lead. Its approximate length is 15 inches due to the velocity factor of the 300 Ω material. The λ/4 piece is shorted at the bottom and thus is an open circuit (high impedance) at the end of the λ/4 section. This matches well to the λ/2 radiator for VHF. The 50 Ω tap is about 1¼ inches from the short, as mentioned before.

For UHF operation, the λ/4 matching stub at VHF is now a 3λ/4 matching stub. This is electrically a λ/4 stub with an additional λ/2 in series. Since the purpose of the matching stub is for impedance matching and not for radiation, it does not directly affect the radiation efficiency of the antenna. It does, however, suffer some transmission loss from the additional λ/2, which would not be needed if it were not for the dual band operation. I estimate this loss at about 0.1 dB. Next comes the λ/2 radiating element for UHF, which is about 12 inches. To

**Table 1**
**Measured Relative Performance of the Dual-band Antenna at 146 MHz**

| *VHF λ/4 GP 4 radials* | *VHF Flexible Antenna* | *Standard VHF J-Pole* | *Dual-Band J-Pole* |
|---|---|---|---|
| 0 dB reference | −5.9 dB | +1.2 dB | +1.2 dB |

**Table 2**
**Measured Relative Performance of the Dual-band Antenna at 445 MHz**

| *UHF λ/4 GP 4 radials* | *UHF Fexible Antenna* | *Standard VHF J-Pole* | *Dual-Band J-Pole* |
|---|---|---|---|
| 0 dB reference | −2.0 dB | −5.5 dB | 0.5 dB |

make it electrically terminate at 12 inches, a λ/4 shorted stub at UHF is constructed using RG-174A. The open end is then connected to the end of the 12 inches of 300 Ω twinlead. The open circuit of this λ/4 coax is only valid at UHF. Also, notice that it is 4½ inches and not 6 inches due to the velocity factor of RG-174A, which is about 0.6.

At the shorted end of the 4½ inch RG-174A is the final 18 inches of 300 Ω twinlead. Thus the 12 inches for the UHF λ/2, the 4½ inches of RG-174A for the decoupling stub at UHF, and the 18 inches of twinlead provide for the λ/2 at 2 meters. The total does not add up to a full 36 inches that you might think. This is because the λ/4 UHF RG-174A shorted stub is inductive at 2 meters, thus slightly shortening the antenna.

## Making it Portable

The single most common question that people asked regarding the DBJ-1 is how it could be made portable. The original DBJ-1 had the antenna inserted into Class 200 PVC pipe that was 6 feet long. This was fine for fixed operation but would hardly be suitable for portable use. Basically the new antenna had to have the ability to be rolled up when not in use and had to be durable enough for use in emergency communications.

The challenge was to transfer the concepts developed for the DBJ-1 and apply them to a durable roll-up portable antenna. After much thought and experimenting, I adopted the configuration shown in Figure 4.

The major challenge was keeping the electrical characteristics the same as the original DBJ-1 but physically constructing it from a continuous piece of 300 Ω twinlead. Any full splices on the twinlead would compromise the durability, so to electrically disconnect sections of the twinlead, I cut small ¼ inch notches to achieve the proper resonances. I left the insulating backbone of the 300 Ω twinlead fully intact. I determined the two notches close to the λ/4 UHF decoupling stub by experiment to give the best SWR and bandwidth.

Because this antenna does not sit inside a dielectric PVC tube, the dimensions are about 5% longer than the original DBJ-1. I used heat shrink tubing to cover and protect the UHF λ/4 decoupling stub and the four ¼ inch notches. Similarly, I protected with heat shrink tubing the RG-174A coax interface to the 300 Ω twinlead. I also attached a small Teflon tie strap to the top of the antenna so that it may be conveniently attached to a nonconductive support string.

Figure 5 shows a picture of the λ/4 UHF matching stub inside the heat shrink tubing. The DBJ-2 can easily fit inside a pouch or a large pocket. It is far less complex than what would be needed for a single band ground plane, yet this antenna will consistently outperform a ground plane using 3 or 4 radials. Setup time is less than a minute.

I've constructed more than a hundred of these antennas. The top of the DBJ-2 is a high impedance point, so objects (even if they are nonmetallic) must be as far away as possible for best performance. The other sensitive points are the open end of the λ/4 VHF matching section and the open end of the λ/4 UHF decoupling stub.

As with any antenna, it works best as high as possible and in the clear. To hoist the antenna, use non-conducting string. Fishing line also works well.

## Measured Results

I measured the DBJ-2 in an open field using an Advantest R3361 Spectrum Analyzer. The results are shown in Table 1. The antenna gives a 7 dB improvement over a flexible antenna at VHF. In actual practice, since the antenna can be mounted higher than the flexible antenna at the end of your handheld, results of +10 dB are not uncommon. This is the electrical equivalent of giving a 4 W handheld a boost to 40 W.

The DBJ-2 performs as predicted on 2 meters. It basically has the same performance as a single band J-pole, which gives about a 1 dB improvement over a λ/4 ground plane antenna. There is no measurable degradation in performance by incorporating the UHF capability into a conventional J-pole.

The DBJ-2's improved performance is apparent at UHF, where it outperforms the single band 2 meter J-pole operating at UHF by about 6 dB. See Table 2. This is *significant*. I have confidence in these measurements since the flexible antenna is about −6 dB from that of the λ/4 ground plane antenna, which agrees well with the literature.

Also notice that at UHF, the loss for the flex antenna is only 2.0 dB, compared to the ground plane. This is because the flexible antenna at UHF is already 6 inches long, which is a quarter wave. So the major difference for the flexible antenna at UHF is the lack of ground radials.

## Summary

I presented how to construct a portable, roll-up dual-band J-pole. I've discussed its basic theory of operation, and have presented experimental results comparing the DBJ-2 to a standard ground plane, a traditional 2 meter J-pole and a flexible antenna. The DBJ-2 antenna is easy to construct, is low cost and is very compact. It should be an asset for ARES applications. It offers significant improvement in both the VHF and UHF bands compared to the stock flexible antenna antenna included with a handheld transceiver.

If you do not have the equipment to construct or tune this antenna at both VHF and UHF, the antenna is available from the author tuned to your desired frequency. Cost is $20. E-mail him for details.

**Notes**

[1]E. Fong, "The DBJ-1: A VHF-UHF Dual-Band J-Pole," *QST*, Feb 2003, pp 38-40.

[2]J. Reynante, "An Easy Dual-Band VHF/UHF Antenna," *QST*, Sep 1994, pp 61-62.

*Ed Fong was first licensed in 1968 as WN6IQN. He later upgraded to Amateur Extra class with his present call of WB6IQN. He obtained BSEE and MSEE degrees from the University of California at Berkeley and his PhD from the University of San Francisco. A Senior Member of the IEEE, he has 8 patents, 24 published papers and a book in the area of communications and integrated circuit design. Presently, he is employed by the University of California at Berkeley teaching graduate classes in RF design and is a Principal Engineer at National Semiconductor, Santa Clara, California working with CMOS analog circuits. You can reach the author at* **edison_fong@hotmail.com**. QST

# A Tilt-Over Antenna Mast Born of Necessity

## A ham's physical limitations lead him to design a convenient mast.

**Bruce Belling, N1BCB**

While I have a capable 55 W home transceiver for 2 meters, it just wasn't enough to reach the repeaters in my area while using an indoor antenna. I live in the woods and in a slight hollow, which necessitated getting my antenna up high enough for line of sight coverage. Because of surgeries I am not allowed on ladders or roofs, so I had to devise a way to construct and raise a mast at least 25 feet high and be able to get it down quickly for service or impending hurricane strength weather.

The antenna I devised will also serve as a termination point for a horizontal loop HF antenna that is strung from tree to tree around my back yard. This antenna is pulled up to about 15 feet using a pulley and small line system. Releasing the line at the mast point allows me to lower the mast without disturbing the loop antenna. A neighbor performed the ladder part of this pulley installation. This horizontal loop plus a dipole antenna gives me two HF antennas that are attached to an antenna switch at my desk. This gives me plenty of HF flexibility.

### Mast Construction Plan

To make the tower, I blocked two 8 foot 2 × 4s together with a enough space between them to clear a 2 × 4. The space was maintained by four 2 × 4 blocks and shims. Then I bolted a third 8 foot 2 × 4 between the two and at the top, so the single 2 × 4 fit 2 feet 8 inches between the two. Then I drilled a ⅜ inch hole 4 inches below the top of the joint and through all three boards. I inserted a ⅜ inch bolt to act as a pivot point (see Figures 1 and 2).

I drilled another ⅜ inch hole 2 feet below the first and inserted another ⅜ inch bolt. This served as the locking pin to keep the assembly together while the mast is raised. At the top of the single board, I fastened two 8 inch pieces of 2 × 4 lumber side by side and perpendicular to the single board. I used this as a backing plate to fasten two pipe brackets that hold the 14 foot long 1½ inch PVC pipe antenna mast. The bottom of the pipe was secured with two 2 × 4 blocks and another pipe clamp (see Figures 3 and 4).

### Pipe Mast

I inserted a 10 foot long piece of wooden dowel up into the PVC pipe and fastened the bottom of it to the single board about 2 feet below the PVC pipe. This was to stiffen the PVC pipe and keep it in line with the 2 × 4 part of the mast when the mast is lowered. You may attach your antennas to the mast in a number of ways, depending on your requirements.

For my application, I put a 90° elbow at the top of the PVC pipe, then a 5 inch piece of PVC, another 90° elbow pointing up and added an 8 inch piece of PVC. The top 8 inch piece of PVC is used to attach my 2 meter J pole antenna. This gets the height up to just over 25 feet. I used the 5 inch piece of pipe to attach the guy lines for stability. Two guy lines ran to the building and one to a tree, giving me a three way triangle pattern and a stable mast in high winds.

In the middle of the main piece of PVC I put a four way fitting so I could add two spreaders. I put a T at the end of each spreader and a 5 foot piece of pipe in each. The reason for this is to provide a base for any vertical antennas I may want to put up in the future. I am in the market for a vertical that will cover 80 and 20 meters, but haven't yet selected one. Figure 4 and the lead photo show how this looks and works.

### Raising and Lowering

To lower my antenna support, I only need to release one line to the building, one line to the tree and the line to the HF loop so that the mast will pivot down without having to take apart anything else. I left enough slack in the cable to allow for this movement. At first I had to raise and lower the mast by hand, which was a little tricky.

I installed a boat trailer winch just above the

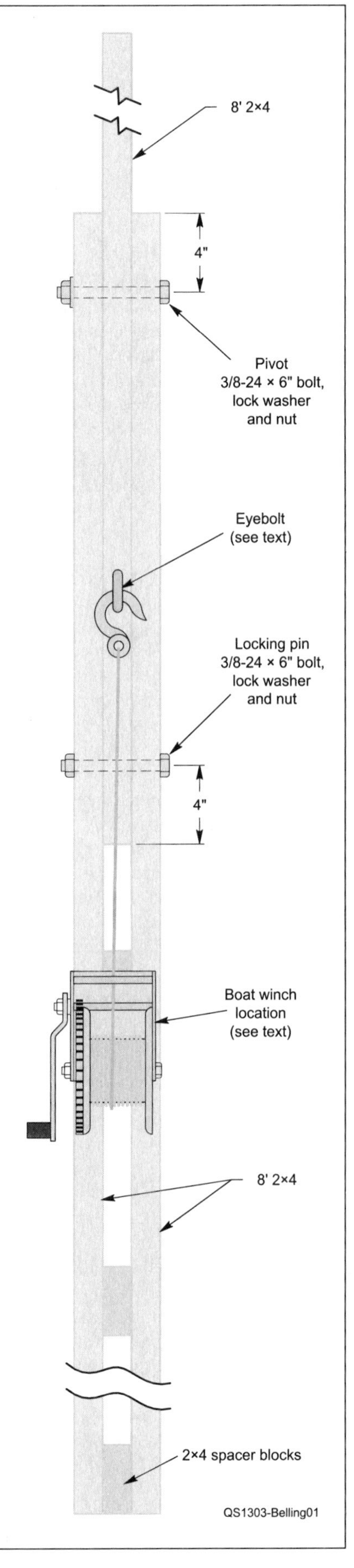


**Figure 1** — Construction details of the lower mast section. The pair of 2 × 4s making the base are secured to the house.

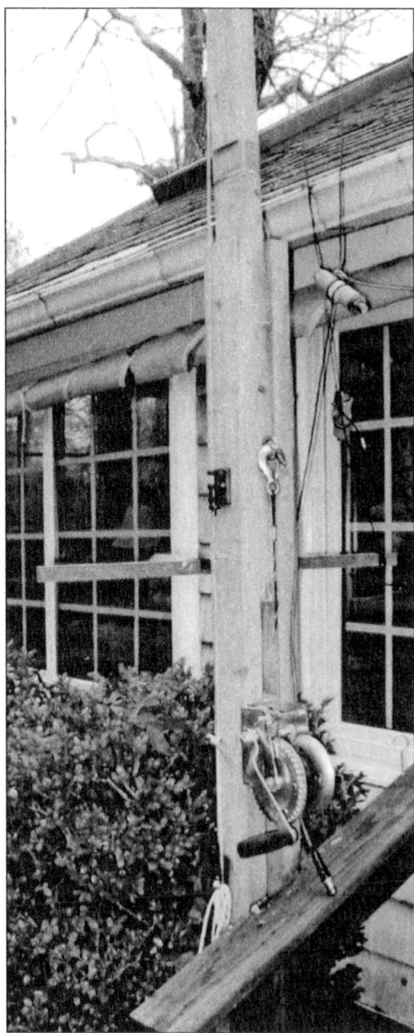

**Figure 2** — Close up of the lower mast section. Note the boat winch, winch cable and snap hook used to safely raise and lower the mast.

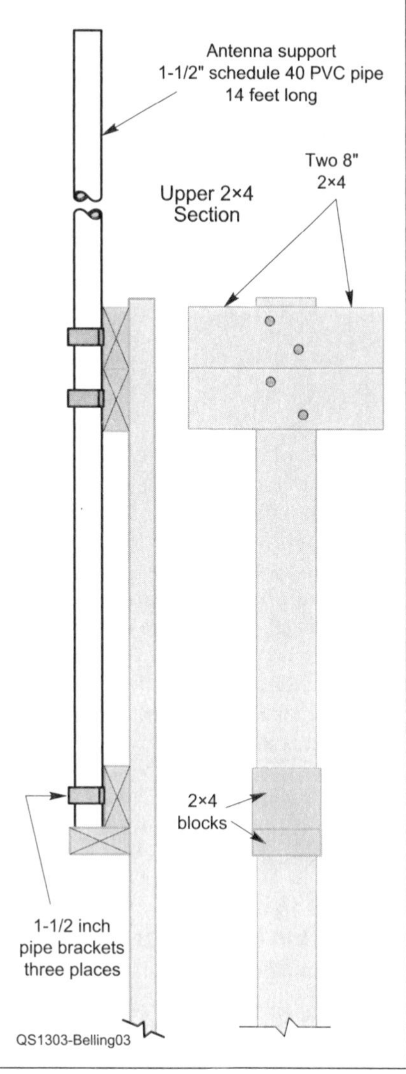


**Figure 3** — Construction details of the upper mast section. The PVC pipe can be adapted to support your antenna systems.

**Figure 4** — Close up of the upper mast section showing the antenna attachment methods employed.

hand rail (see Figure 2) and attached the cable to the butt end of the upper 2 × 4. Now all I have to do is loosen the cable a small amount, lock the winch, remove the locking pin and push the mast over enough to offset the balance. Then I unlock the winch and lower the mast. I find that this takes 10 minutes or less, and about the same to raise and tie and lock everything in place.

If there are any storm warnings indicating high winds, I drop the mast which, when in horizontal position, is only about 6 feet off the ground and 3 feet from the house. So far, the mast has been fine in winds up to 40 mph. I wouldn't worry too much about higher wind velocities but I would lower the mast if hurricane warnings came in.

Perhaps there are others out there with physical limitations such as mine who could use a plan like this, or would appreciate the convenience.

ARRL member and General class licensee Bruce Belling, N1BCB, has been licensed since 1995, but considers himself new to Amateur Radio, since he didn't buy his first radio until 2008. You can reach Bruce at 113 Brayton Point Rd, Westport, MA 02790 or at **jabrwest@charter.net**.

# A One Person, Safe, Portable and Easy to Erect Antenna Mast

*Consider this approach to Field Day antenna installation — especially if you're short of trees!*

**Bob Dixon, W8ERD**

Widely available military surplus mast sections can be used in a variety of ways to make excellent amateur antenna masts. The key to making them work well, in my experience, is the mast tripod available from Barans (**stores.shop.ebay.com/barans-military-surplus-and-radio**). The tripod consists of three downward angled sockets that the mast sections plug into, and a vertical central shaft that the mast sections can slide through. This makes four contact points on the ground, providing a stable base. Note that this is different from the military GRA-4 tripod, which does not work for this purpose. Figure 1 illustrates the tripod and masts, and shows the general idea of how they all fit together.

Many of us have had bad experiences with push-up type masts that are difficult to erect, even with a crew of several people, regardless of whether you stand on a ladder and push each section up, or use the tilting up technique. Attempting to raise these masts by tilting often breaks them off in the middle. Either approach can be dangerous for the erectors. The mast erection system described here eliminates all those problems.

The military surplus mast sections are often available at hamfests and on the Internet (see Table 1). Their original purpose was to hold up camouflage netting. Each section is about 4 feet long, and the aluminum ones are about 1¾ inch outside diameter and 1½ inch inside. I don't recommend any other size. They are also available in fiberglass, but be careful if you buy those. Many of them are defective rejects and break easily at the joints. They must have the reinforcing ring at the ends. For this mast project, fiberglass sections can be used for the tripod lower legs if you wish, but they are not suitable for the vertical portion because they will not fit thru the tripod.

## Assembling and Erecting the Mast

One example of a 40 foot mast that can be constructed with these materials is explained here in detail, although many other combinations are possible, and are limited only by your imagination. See Figure 2 for the general concept. You need six mast sections for the legs. The vertical part is 10 sections high, and must be aluminum. You will need a total of 16 mast sections.

You will also need guy ropes, a guy ring (see Figure 3) and a mast clamp (see Figure 4) to fasten the ropes to the masts. If you plan to use the mast to hold up a wire antenna, mount a pulley and pull rope (halyard) at the top, so you can pull up the antenna after the mast is erected. Snap rings are also available to make it easier to attach the guy ropes to the guy rings, mast clamp, guy ring and ground stakes.

The erection process is what makes this design so nice. Make sure you have a level surface to start with. Start with one mast section and plug it into a tripod angled socket. Continue around the tripod with two more mast sections and set the initial tripod structure upright. Then start with one leg, tilt the structure backwards, and add another mast section. Again continue this around the base until you have two mast sections in each leg. Now put two mast sections up the middle. This puts the initial top of the structure at a height that is easy to work with, while adding things to the mast. Add the top guy ring (with loose guy ropes attached) and the pulley (with a loose loop of rope equal to at least the height of the mast, threaded through the pulley) to the top. Or attach a small antenna and coax to the top if desired. Then grasp the center mast below the tripod, lift it up about 4 feet and insert another mast section. Slide more sections up from the bottom, one at a time. One person can do this, and it will easily support itself to 40 feet, as long as it is level and there is little wind.

When you reach the halfway point, install the mast clamp, along with its loose guy ropes. Note that a normal guy ring will not work there, because you can't get it

**Figure 1 — Lower portion of the mast, with my wife Judy and dog Olivia for scale.**

**Table 1**

**Some Suppliers of Masts and Accessories**

Barans Surplus, **baranoskybunch@aol.com**

Bayway Deals, **stores.shop.ebay.com/bayway-deals**

TeeVee Supply, **www.teeveesupply.com/product_pages/antennas/antenna_mounting_hardware.htm**

The Mast Company, **www.tmastco.com**

Note that the dealers do not always list all their materials on their web pages. If necessary, send them e-mail and inquire.

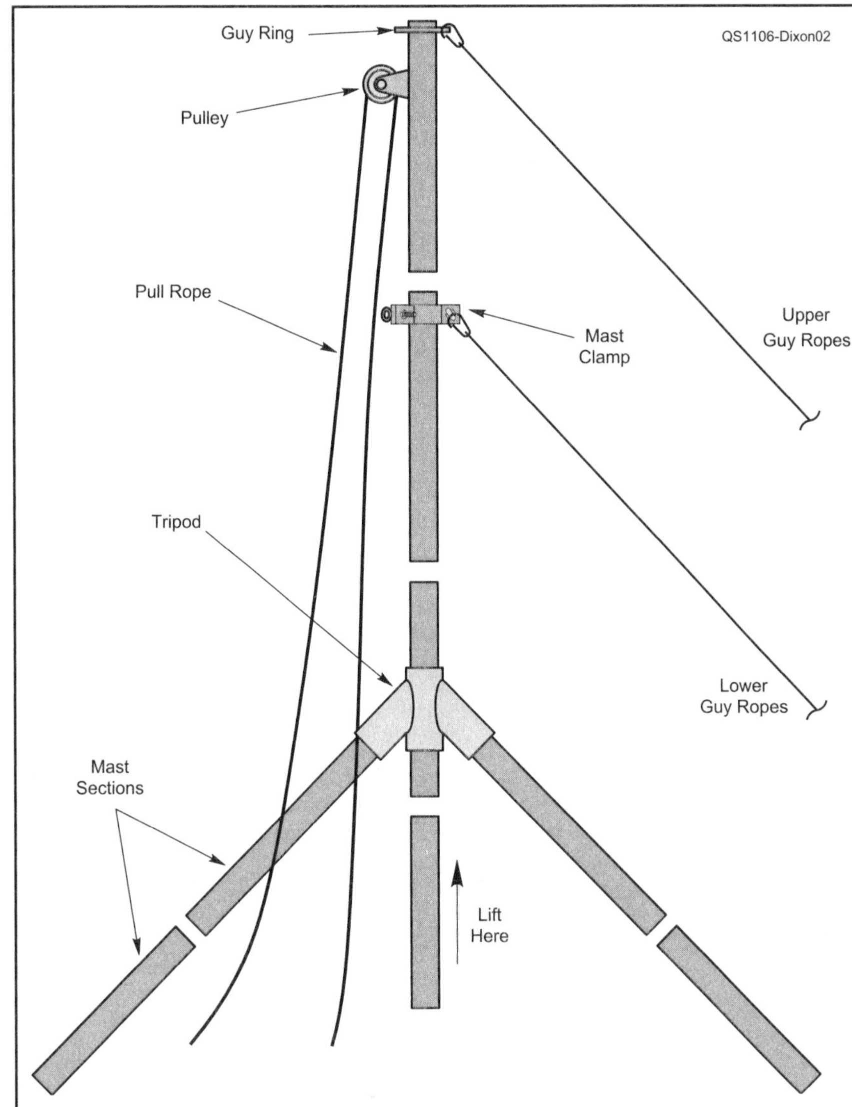


**Figure 2 — Portable mast functional diagram — not to scale.**

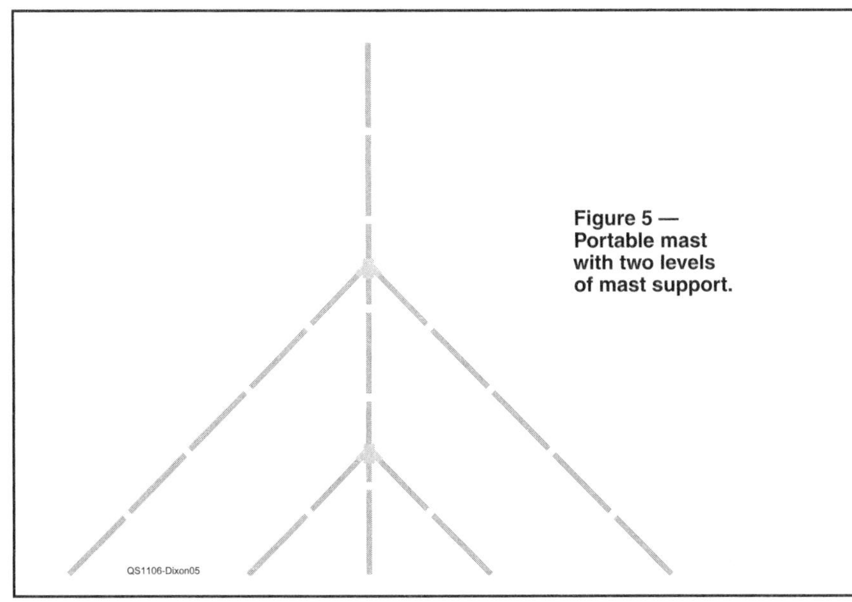


**Figure 5 — Portable mast with two levels of mast support.**

**Figure 3 — Guy ring used to fasten guy ropes to the top section of the mast.**

**Figure 4 — Mast clamp used to fasten guy ropes at intermediate heights.**

onto the mast once you start the erection process.

When the mast is fully up, pound in the guy rope stakes at suitable locations, and tighten all the guy ropes evenly so the mast stands straight. Attach a wire antenna to the pulley rope and pull it up. A video of the installation process is available on the QST-in-Depth website.[1]

## Disassembly and Storage of the Mast

To take it down, just reverse the above process. It slides down very easily. Start by detaching all the guy ropes from the guy stakes. Leave the other ends of the ropes attached to the guy ring and mast clamp permanently.

After the mast is completely down and disassembled, wind each rope up separately starting from the inner end, and tape each in two places. Then tape all the rope loops together at each level. Next, tape both levels of rope together to leave you with a single coil of tangle free guy ropes for the next time. Also wind up and tape the pulley rope and leave it on the pulley.

The mast sections usually come with canvas storage bags. I have found that home supply stores offer a medium size latching

[1]**www.arrl.org/qst-in-depth**

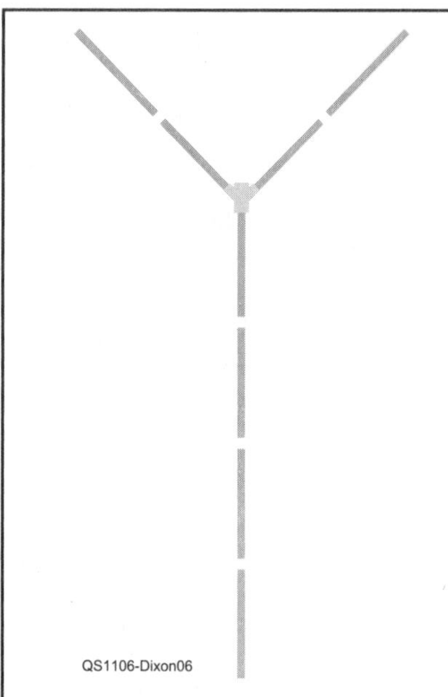


**Figure 6 — Mast with inverted tripod and top loading for a vertical antenna.**

plastic storage bin that perfectly holds all the other parts. The Home Depot store calls theirs Sterilite.

Keep in mind that gravity and the guy ropes are what holds all the mast sections and the tripod together, so if there is a strong upward wind or the guy ropes come off, the sections could separate. These masts can also be used nicely for fixed antennas, and in that case you should drill holes and insert bolts at each joint.

The Delaware Amateur Radio Association (Delaware County, Ohio) used two of these masts with great success at the last ARRL Field Day.

## Other Possibilities

Other mast configurations are possible, and some are illustrated in Figures 5 and 6. You could have more than one level of "guy masts," for a strong higher mast. Or you could invert a tripod and put it at the top and make a top-loaded vertical antenna. In that case, drill small holes in each top mast section and use tap screws and wire to be sure there is good contact between the masts and the tripod. The masts themselves make good electrical contact with one another, but the tripod may not make contact through the paint on the mast sections.

This mast is also very useful as a temporary mounting for an antenna, such as a VHF or small HF Yagi that is being assembled, tested and tuned. It will easily support such antennas at a low working height for assembly and initial testing, and can then be temporarily extended to 16 or 20 feet to determine the effect of height on tuning.

*Photos by the author.*

*ARRL Life Member and Amateur Extra class operator Bob Dixon, W8ERD, was first licensed in 1955 as WN9OKN, and then progressed to W9OKN before receiving his current call. He prefers operating CW on the HF bands and 6 meters, and is active with the Delaware County (Ohio) Amateur Radio Association and ARES®. He has confirmed every DXCC entity except North Korea. He is partially retired, but continues to work in the field of computers for the Ohio Academic Resources Network, where he has worked with video conferencing, satellite Internet and providing Internet services to tiny Appalachian towns. He received BS and MS degrees in Electrical Engineering from the University of Wisconsin, and a PhD from Ohio State University, working with John Kraus, W8JK. He is a Senior Life member of IEEE, and a licensed Professional Engineer. He has worked in the field of SETI for many years at Ohio State, and now with the North American Astrophysical Observatory, developing a new omnidirectional radio telescope concept called Argus (***www.naapo.org***). You can reach Bob at 2131 Klondike Rd, Delaware, OH 43015 or at* **w8erd@hughes.net**.

# The Fiberpole — Evolution of an Antenna Mast

**Dave Reynolds, KE7QF**

Build this lightweight mast for use in fixed or portable applications.


Over about 30 years, I've overcome the challenges of using flexible antenna masts, inspired by a fascination with the fiberglass pipe that's used at underground at gasoline service stations. It's classy stuff — 300 PSI burst strength but a weight of less than 10 pounds for a 20 foot length. The result is an antenna that you can put up and take down in a matter of minutes all by yourself. I liked mine so much that I gave it a name — Fiberpole.

You can replicate such a mast with beginner's skills and ordinary tools. Neighbors were amused as I first erected what looked like spaghetti, bending back on itself, with me pushing it upright with my 6 foot height. Once secured, the stability was great, the coax stayed dry and it wasn't very ugly.

My Fiberpole, together with its gin pole, has gone to ARRL Field Day atop my car, disassembled and folded upon itself, elements, coax and all, wrapped in a blanket, and returned to get back on the air by Sunday night. Now that's portable!

## Making the Mast

My original mast was assembled from two 20 foot lengths of pipe spliced in the center. Strength for center lifting was improved, however, when two nearly 10 foot sections were spliced, one onto each end of a 20 foot center section. This also improved access to the ends for antenna and cable exit details. My splices ended up being 10 inch lengths of the pipe, cut lengthwise to provide a 9/16 inch lengthwise gap. This gap gets compressed and the splice fits snugly into the inside of the pipe. So now I'm a bit short of 39 feet.

File or sand the cuts to remove burrs, and then test for a compressed fit inside the mast. There needs to be a small gap in the installed splice. Fiberglass is easy to drill, saw, file and even cut with a knife, but it dulls cutting tools quickly. Don't use your favorite miter saw. A metal cutting hacksaw with its disposable blade, or an electric saber saw works well. To mark for square cuts, wrap straight-edged cardboard around the tube and draw along the resulting square edge. I mark lengthwise for the long cuts with a small piece of angle aluminum, which results in good parallel lines.

Avoid breathing dust from this, as with any cutting operation. Use appropriate eye and ear protection with power tools. Mechanics' gloves from an auto parts store can reduce hand abrasions.

Cut off both tapered ends of the remaining 20 foot length of pipe; it'll require removing about 3 inches from each end. This finishes the center section, and now you're ready to assemble the end pieces onto this long center section. Assembly is easier if you compress the middle of the splices with a radiator hose clamp and insert it to join the mast sections, then remove the clamp and push the sections together.

**Figure 1 — Assembly of sections over the splice before removing the clamp. Note the centering screw at the top of photo, and the reassembly alignment arrows.**

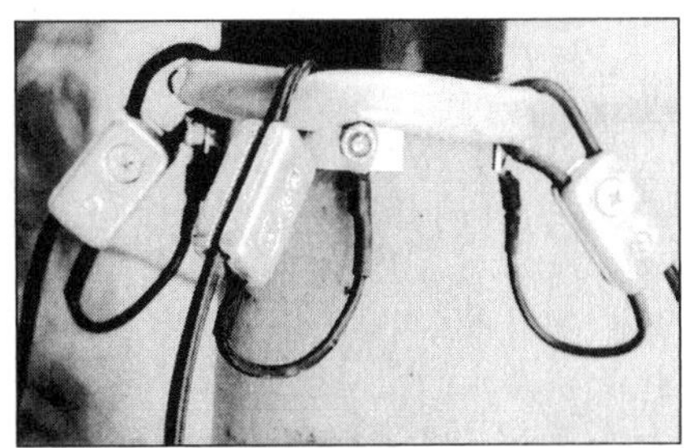

**Figure 2 — Flexible attachment ring with wires attached to the screws.**

Figure 1 shows an upper part of the mast that is to be assembled over the compressed splice extending from within the lower section. Note the arrow marks showing the correct alignment of the sections for reassembly. I have a short, blunt-point self-tapping screw through the outside and into one half of the splice to keeps things from migrating. You might try an adhesive applied to one half of the splice.

Connect the feed line on the inside of the top section to the antenna elements on the outside. I use a 2 inch diameter by ⅛ inch thick plastic bulkhead with an SO-239 coax jack, threaded-end-down at the center. The coax hangs from this connection, so assemble it securely. I made a ring about 4 inches in diameter of some cable sheath with several turns of nylon rope inside, and the elements now pull against each other through that flexible attachment ring as shown in Figure 2. The elements are looped over the torque ring, fastened with a cable clamp, and a short length of flexible wire connects them to the screws. Several changes of RG-213 coax have stayed snug and dry inside the mast these many years in Colorado, Ohio, Canada, Arizona and now at my mobile home in Wyoming.

## Raising the Mast

The essential items are a gin pole, a base hinge plate, and an upper locating method.

### *The Gin Pole*

The Fiberpole all came together with the addition of a gin pole. With this, I was able to lift the mast at its midpoint instead of trying to push it up. My gin pole is a 20 foot long steel pipe with a pulley at the top. The gin pole is firmly positioned vertically. A haul rope then goes through the pulley at the top to lift the mast upward by its middle. You stand at ground level beside the gin pole and pull downward on the haul rope and the mast raises with perhaps 25 pounds of pull required. A good source of galvanized steel pipe is the typically 20 foot top rail used for chain link fences.

### The Hinge Plate

The hinge plate at the bottom simply holds the bottom of the mast in place and keeps it there as you raise it from the ground. As a result of using a gin pole, you pull just over 20 pounds of force straight down as you raise Fiberpole with a center guy rope. Level the hinge plate to reduce hinge pin stress. My hinge plate is wood, but still over-engineered; probably a ¼ inch bolt or a big cotter pin would do nicely, considering the light weight. Let your drills and junk box make the decision for you, but don't try to do without a base hinge fixture such as shown in Figure 3.

### The Pulley

You need a smooth running pulley at the top of the gin pole, and a metal loop over the pulley to keep the haul-rope from escaping over the side. A small cleat at eye-level on the gin pole is handy for tying off the haul-rope when the mast reaches vertical.

Assemble the Fiberpole and its elements on the ground, and attach it to the hinge plate. I have a ground rod coming through a hole near the center of my plate and the ground wire clamp holds the plate down. The plate needs to be free to rotate slightly to allow the lowered mast to be shifted.

Locate the guy ropes, if you use them, around the mast at gin-pole height, free to center themselves through a wire loop secured by a hose clamp.

The raised mast is placed into a socket fixture, which needs to be perhaps 8 or 10 feet directly above the base. Mine is plywood, attached with an angle steel bracket near the roof of my mobile home.

### Pulling it Up

My upper locating socket has a deep notch opening toward the direction to which the mast lays down. The raised mast is confined within this socket. When you've hoisted the Fiberpole into position in the upper locating socket, tie off your haul rope on the gin pole cleat. The gin pole in its notch in the upper locating socket will hold the mast secure while you and sort out the elements dangling around you. On the side of the socket fixture opposite the mast, there's another notch for the gin pole (which in my case is 1⅜ inch outside diameter steel). The gin pole needs no lock wire.

Directly below the gin pole notch, protruding upward from the hinge plate, is a ⅜ inch machine bolt standing about an inch high. The gin pole fits loosely over this bolt to position the bottom of the pole. If you think your gin pole is too flexible, you can strengthen it with a guy rope to the rear.

**Figure 3 — Wooden hinge plate using a surplus TV dish bracket.**

**Figure 4 — Close view of the Fiberpole in mid lift. The gin pole is on the left.**

During the erection process (see Figure 4) it's important to lay the elements and ropes out in their correct directions and order on the ground, so that they end up on the proper side of the socket and rope. With the gin pole in place, haul the mast up to vertical and into its side of the socket fixture. Secure the haul rope around the cleat on the gin pole. Recheck to be sure all the elements and guys will end up in the directions you intend for them.

Secure any guy ropes that are used, connect your coax and trim the elements to length, guided by SWR readings at several nearby frequencies to determine the changes in length per kilohertz that you require.

## Finding the Pipe

This fiberglass pipe comes in 2 inch and 3 inch inside diameters. To find a source, check the phone directory under *Service Station Equipment and Supplies*, or talk to someone in the gasoline service station repair business. You can expect this pipe to cost between $4 and $7 a foot. One brand name of this pipe is Ameron.

## Closing Notes

There was a noise problem from the cable slapping against the inside when the mast flexed in the wind. I cut some pieces of carpet and secured a couple of these along the length of the cable, and the noise stopped. The exit for the coax cable can be out the bottom or at roof height through a ¾ inch hole in the mast.

If you're storing this pipe for long periods, it's best to support it so that it lies flat, and preferably out of the sunlight. If one side deteriorates more than the other, it results in a slight bend. I suggest painting the Fiberpole to protect the epoxy resin from ultraviolet deterioration that results in fuzzy glass fibers on the surface. The best coating I've found is a latex primer tinted to a gunmetal gray (neighbor friendly), applied with a brush and covered with a similar finish coat of paint.

*ARRL member Dave Reynolds, KE7QF, was first licensed in 1972 as a Technician with the call WBØDXJ. He was later WBØLNZ, as a General and Advanced class licensee and received his present call in 1983. He's written articles for the* 1995 *and* 2000 ARRL Handbooks *about circuit board construction and the use of surface mount electronic parts. Dave is active in clubs and nets and has served as chairman of the Amateur Radio Council of Arizona. You can contact him at PO Box 538, Evansville, WY 82636 or* **daveke7qf@msn.com**. QST

# A Vehicle Mounted Mast for Field Operating

*This inexpensive mast gets you extra range from your parked vehicle without an extra footprint.*

**Geoff Haines, N1GY**

This project actually came about because of a little visit to one of those charity operated thrift shops that are very common in many parts of the country. After dropping off our donations to the cause, I was just scanning the room when my eyes lit on a bright yellow object in a corner. It turned out to be one of those telescopic poles with a spring loaded light bulb changer basket on top. The less than $5 price was irresistible, so I paid the bill and took it home. The project was already taking shape in my head.

I have always liked to be able to operate from my car at a park or beach. The only problem with setting up my big telescopic mast, as described in a *QST* article, was the need to set up the base, then drive the car onto the base, then mount the mast, extend it, guy it down and only then be able to hook up the coax and begin to operate.[1] While this can be a one person job, it is much easier with two or more. The guying also takes up a fair amount of room around the car, which can be a problem.

## A New Approach

This new mast would have to be designed to be totally contained within the footprint of the car. No guy ropes, no separate base — just erect the mast and antenna and get on the air. My SUV already has a trailer hitch stud on the rear bumper. A length of 1 inch inside diameter PVC pipe fit both the hitch stud and the lower section of the light bulb changer mast perfectly (see Figure 1). The next job would be to convert the changer basket to something that could accommodate a V/UHF antenna such as the one I described in August 2006 *QST*.[2]

Amazingly, once I removed the light bulb changing basket's components, I was left with a screw-on base that, after a little work with a narrow chisel, was ready to be mated with the PVC T at the base of my antenna. Four short sheet metal screws were used to secure the screw-on base to the antenna. Now the antenna can be simply screwed onto the top of the mast and secured with a small setscrew that was included with the base.

[1]Notes appear on page 110.

## Mounting the Mast to the Vehicle

The next task was to ensure that the mast, which in its stock form extends about 11 feet, was rigidly mounted and stable. This is important because once the PVC pipe is added to the bottom of the telescoping mast and the antenna is added to the top of the mast, the whole assembly is 21 feet from the ground to the top of the antenna. That ought to add some range compared to the magnetic mount antenna on top of the car's roof.

In looking around the SUV, I discovered that the adjustable crossbar at the rear of the rooftop rack has tie down loops at each end. Some scrap ½ inch diameter aluminum tubing that was left over from some long dead antenna provided enough material for two braces that run from those tie down loops to a pipe clamp attached to the PVC pipe. The ends of the aluminum tubing were flattened and bent to the appropriate angle. Holes were drilled in each end to accommodate stainless steel bolts with self locking nuts at the ends that go into the tie downs (see Figure 2) and a ⅜ inch stud attached to the pipe clamp (see Figure 3). A ⅜ inch wing nut secures the braces to the pipe clamp and the bolts at the other end of the braces slip into the tie-down rings. These ends are secured with a bungee cord across the luggage rack and the braces, exerting downward pressure to keep the braces in position. Another bungee cord is run from the wing nut and stud down under the rear bumper to again apply downward pressure to keep the mast assembly in position.

The length of the aluminum tubing braces was carefully measured after experimentation to ensure that the mast assembly is truly vertical in all planes when erected on a level parking spot. Adjustment fore and aft is achieved by sliding the pipe clamp up or down the PVC pipe and side to side adjustment by twisting the pipe clamp left or right before securing the clamp to the pipe. These adjustments are relatively narrow in scale, so careful measurement of the length of the braces is vital. Once all the adjustments are completed, the setup is done.

**Figure 1 — The mast setup without an antenna.**

## Taking it Down

Disassembly and removal of the mast and braces should not result in having to readjust again. In setting the system up, once a suitable parking location is reached, one simply assembles the braces and mast (with the antenna mounted) and the mast will always be vertical if the vehicle is level. Only if the parking spot is not flat and level will more adjustment be required. The entire system breaks down into seven pieces:

- One 1 inch inside diameter PVC pipe.
- One telescopic mast.

■ Two aluminum braces.

■ Two bungee cords.

■ A dual band antenna.

All of this fits easily into the back of my SUV along with my emergency go-kit and the other stuff I have accumulated. Using the smaller mast means that the setup is completely contained within the space of the vehicle. This makes setting up and breaking down much easier and takes less time.

## Wrapping Up

There are a couple of caveats that come with this project. First, under no circumstances should one try to drive the vehicle with the mast erected. Contact with overhead power lines could be fatal. Second, because of the light duty nature of the mast, only relatively small and light VHF or UHF antennas are suitable. Larger antenna arrays need larger and more rugged mounting systems to be safe.

My costs were minimal because I already had the PVC pipe and the aluminum tubing on hand. Besides the cost of the telescopic light bulb changer, which should be available at any home improvement store, I spent another $10 or $11 on stainless steel fasteners and the pipe clamp. The contents of your junk box may vary but I doubt that you could spend more than $50 on the project even if everything had to be purchased new.

This might make a good first project for a newer ham. The extra range that a 21 foot high antenna gives the operator over the usual mobile antenna will come in very handy, particularly if they get involved (as they should) with their local ARES® group or public service event support.

**Notes**

[1]G. Haines, N1GY, "The Octopus — Four Band HF Antenna for Portable Use," *QST*, Dec 2007, pp 36-38.

[2]G. Haines, N1GY, "A Neat Dual Band Antenna," *QST*, Aug 2006, pp 50-51.

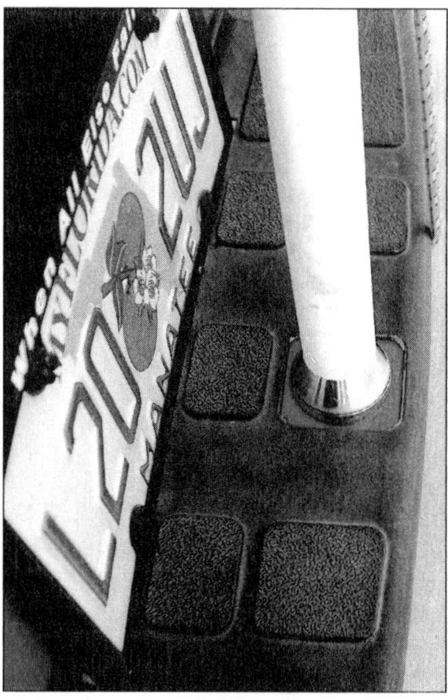

**Figure 2 — The bottom end of the PVC pipe slips over my trailer hitch stud. It could easily be a regular bolt and nut through the bumper if one does not have a tow hitch on the vehicle.**

**Figure 3 — This detail shows the end of one of the braces, with the end slipped into a tie down loop on the luggage rack. The bungee cord exerts downward force on the brace to keep it securely in the tie-down loop.**

**Figure 4 — The braces attached to the PVC pipe via the pipe clamp. The bungee cord connects to the rear bumper of the vehicle to exert downward force keeping the mount in place.**

*Photos by the author.*

*ARRL member Geoff Haines, N1GY, was first licensed in 1992 as N1LGI. Geoff upgraded to Amateur Extra class in 2005 and obtained his current call sign. He retired after a career in respiratory care. Geoff currently holds several ARRL appointments in the West Central Florida Section, including Assistant Section Manager, Technical Coordinator and Net Manager among others. Geoff is President of the West Central Florida Group, operators of the NI4CE repeater system, a past president of the Manatee Amateur Radio Club, and a member of several ham radio clubs both in Florida and Connecticut. In his spare time, Geoff is the Editor of the quarterly e-magazine* The Experimenter *for the West Central Florida Section. Geoff is active in designing small projects such as antennas and accessories suitable for the new ham He also finds time to update his Web site:* **www.n1gy.com** *on a regular basis. Geoff can be reached at 904 52nd Avenue Blvd, W Bradenton, FL 34207, or* **n1gy@arrl.net**. QST

# A Portable Antenna Mast and Support for Your RV

*This antenna support can help you get on the air quickly from your recreational vehicle.*

**Paul Voorhees, W7PV**

One of the big frustrations with setting up antennas in a campground or in the field for contesting is getting the thing up in the air without getting a hernia or needing the help of two other adults and three children who would rather be doing something else.

Since there is something enticing to me about operating in the field, either contesting or just hanging out in a state park, I really wanted something simple and cheap that I could attach to my RV to get my 2 element 20 meter beam up high enough to use it. Most of the commercial products I found were expensive and even more objectionably required guys and supports that were potential hazards to other park users.

### A Solution Appears

While eyeballing my RV trying to figure out where to attach some mast supports without drilling a lot of potentially leaky holes, or otherwise attaching something functional but ugly, I noted that the 4 inch square bumper (that most RVs have for storing their plastic sewer hose) would be an almost perfect fit for a piece of 4 × 4 inch fence post. I could then attach my antenna support to the 4 × 4. I had previously purchased several 4 foot sections of surplus fiberglass mast on an Internet auction site. I have seen similar items at several hamfests for just a few dollars per section.

Next I headed for my local hardware store to figure out how to attach the mast sections to the 4 × 4. I took a section of mast along to try to make sure this was a one trip adventure. I did get a weird look after responding to the usual question from the person in the orange apron, "Are you finding everything okay?" My response "I'll know when I see it."

**Figure 1 — The flange is screwed onto the end of a 24 inch 4 × 4, as shown.**

I came away with some ¾ inch diameter galvanized pipe pieces: a mounting flange, a 2 inch nipple, a 12 inch nipple and two double female pipe connectors. The pipe connectors were chosen as they had just the outer diameter I needed for a good fit to the inner diameter of the mast. This all came to about $13. Obviously, if you could find a pipe with the right outer diameter (and a flange to go with it), you could avoid purchasing more pieces. As you can see from Figure 1, the flange is screwed onto the end of a 24 inch 4 × 4, and the pipes and two connectors are attached as shown.

I recommend drilling some small holes through the connected fiberglass sections to allow a piece of stiff wire, about 4 inches

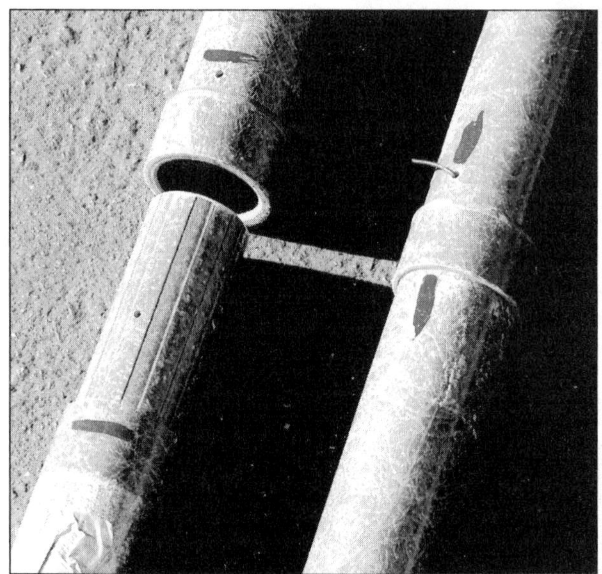

**Figure 2 — View of the mating ends of the fiberglass mast sections. The marking is needed because each joint's locking wire position will be slightly different.**

**Figure 3 — The 4 × 4 is positioned in the 4 inch square bumper, used for storing the sewer hose. Shims are used to secure it as described in the text.**

long, to be fed through and bent over to keep the mast sections together (see Figure 2) while mounting and turning. It's also advisable to dab a couple drops of paint on each connecting pair as shown, using different colors or marks, to make sure the small holes match up for future installations. After disassembly, I just duct taped the wire to the mast sections so it's always there.

To snug the 4 × 4 in the bumper, I added a couple of shims tacked and glued to the bottom. Also, to get the mast properly vertical with the RV level, I similarly shimmed the base of the flange with some scrap sheet metal. The installed mount is shown in Figure 3.

## Getting It Up

To get the antenna in the air, I first assemble the beam on a picnic table, attach the coax, and then put the four mast sections together and lock them using the short pieces of wire mentioned earlier. I use a two element Superantenna, but a Spider Beam or other lightweight array could be used. Then the antenna is tilted up off the table and attached to the tapered end of the mast. At this point I can lift the whole thing by grasping the bottom two mast sections. Holding it vertically, I then lift it over the mount on the bumper. With the mast, it only weighs about 16 pounds. The antenna is then about 19 feet off the ground, but if you want to use another mast section to get to about 23 feet, it is likely that a ladder would be needed, or someone standing on the bumper to help steady it while it was lifted into place.

So, for about $25, I have a self supporting mast and base that is light weight, can be erected by one person in just minutes and requires no guys. It easily supports my two element beam, and could probably handle something even larger, or be the center support for a wire antenna. And, to the relief of my wife, once removed it leaves no trace of having ever been there.

If you don't have an RV, you could probably use the same basic approach and screw or bolt the flange to a piece of 2 × 6 inch lumber, and park a car wheel on top of it.

To try out my new creation I headed over to Lake Wenatchee State Park in eastern Washington. I found a campsite with enough area free of trees so I could rotate my beam. I received some signal reports from distant stations at well over S9 on 20 meters with 100 W. It just doesn't get any better!

*Photos by the author.*

*ARRL member and Amateur Extra class operator Paul Voorhees, W7PV, was first licensed in 1963. He has held calls K7YHQ, W7CHF, WØPAB and W7KXM. After 2 years in the military, he spent 31 years as a financial manager with the US government, living in the Seattle/Bremerton area, St Louis, northern Virginia and Albuquerque. Paul retired in 2003 to his current location.*

*He enjoys travel, grandkids, volunteer work in health care, camping with his travel trailer, and of course, hamming whenever and wherever possible. He has operated from Korea, Johnston Atoll, the Virgin Islands, Western Europe and South Africa. He is currently exploring satellite radio, PSK31 and remote HF operating.*

*You can reach Paul at 10090 Misery Point Rd NW, Seabeck, WA 98380 or at* **ropavo@wavecable.com**.

# Amateur Use of Telescoping Masts

## Do you need an inexpensive skyhook for your antenna? Here's one candidate you may have overlooked.

By R. P. Haviland, W4MB
1035 Green Acres Cr, N
Daytona Beach, FL 32119

One of the many pieces of equipment designed for the TV industry is the telescoping mast used to support TV antennas. In some areas more than about 20 miles from a TV transmitter (and not served by a cable TV company), these masts are used frequently.

TV telescoping masts aren't commonly used by hams. Observation and inquiry show two reasons for this: (1) a lack of knowledge of the capabilities and the limitations imposed by telescoping-mast design; (2) having had (or heard of) a bad experience with such a mast. The latter seem to stem totally from mishandling during installation, leading to mast and antenna damage and—occasionally—injuries to people. The mishandling is directly related to a lack of knowledge regarding proper installation techniques.

### Mast Design Limitations

It's important to know the limitations imposed by telescoping TV mast design. One major TV mast manufacturer's flier emphasizes this by stating "Telescoping masts are not recommended for commercial or ham installations" in the specification list. The "not recommended" ignores the fact that many amateur antennas are *smaller* than some TV antennas used in rural areas and that the mast alone can be used as a vertical antenna.

None of the ads, fliers or instructions packed with a telescoping mast provide adequate information about the capabilities and limitations of its design. The most I've seen is a short note to the effect that the mast should not be used with antennas having a wind load area of more than 2 square feet. In TV antenna ads, you're told the number of elements a particular antenna has, but not necessarily its wind load area or its weight. In most amateur antenna ads, this vital information is given. Telescoping-mast fliers and instructions also omit correct installation procedures and safety procedures. About the only precaution I've seen mentioned is to "stay away from power lines."

Here, I'll explore the capabilities of telescoping masts for amateur use. I'll provide data to allow you to determine correct use in your application, up to the limit of mast capability. Proper installation and use are also covered.

### Telescoping-Mast Design Principles

A telescoping TV mast is much more than a few pieces of steel tubing. It includes a number of design features to make use simple and safe. These feature areas are shown in Figure 1, starting at the top of a mast section and working toward the bottom.

There's size reduction in the top section to bring its ID to a size slightly greater than the next smaller section, which is also the next higher section when the mast is extended. Partly, this is for strength, to distribute the load from the upper section. It's also an antirattle feature, to limit noise caused by wind moving the mast.

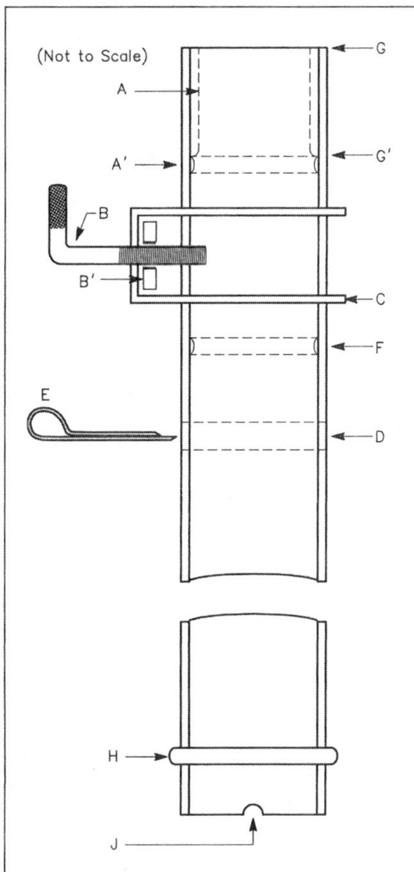


Figure 1—Major features of a single section of a typical TV telescoping mast section. See text for explanation of elements A through J.

In some designs, the top 4 to 6 inches of a section is roll-swaged to the smaller diameter as at A. In others, the swage may extend only 1/4 to 3/8 inch, as shown in Figure 1 at A'. In these, the swage is typically 1 to 2 inches below the top of the section.

Just below the swaged area are two holes in the section. The top one (B) is typically 1/4 inch in diameter, and penetrates only *one side* of the tubing. This hole is for a clamping screw. The screw's primary use is to hold the smaller, inner-mast section in place while a new grip on this section is taken during mast erection. Secondly, the screw keeps the two sections tightly together to prevent wind noise. This screw is *not intended* to be the permanent—and only—mechanism to prevent slippage and mast collapse (more on that later).

The clamping screw (B) has some type of a support fixture. One style uses a simple strap, with a floating square nut for the mating screw. The nut is held in place by metal tabs. A second type (C) uses a flat, **U**-shaped metal sheet, with a hole on each of the **U**-shaped sides just larger than the tubing diameter. There's a captive nut (B') at the bottom of the **U**. Either type may have a small, internally threaded stud not shown here. This stud is used to mount an insulated standoff for carrying twin lead or coax, holding it away from the mast to reduce signal loss and wind-induced slapping noise.

The second, lower hole (D) is typically 5/16 inch in diameter, and penetrates the tube completely, along the diameter. This hole accepts a large cotter pin (E), which is the support for the next-smaller mast section. (The best designs equip the cotter pin with a short chain to prevent the pin from being lost.) This pin is intended to carry *all of the weight above it:* tubing sections, rotator, antenna and guys. It's a safety feature: With overload, the pin should shear and the mast should safely telescope downward. As we'll see, this is the weak point in the design.

In designs I've seen, there's an additional, smaller-diameter swaging (F) between the two holes. The contact of this and the upper swage create a force couple, which transfers bending forces from the smaller section to the larger. It's also part of the antirattle design. Additionally, it's a safety feature (working with the next feature to be described) to prevent sliding the small section completely out of the larger one while raising the mast.

Rings are employed for attaching guy wires. Some designs make the hole diameter in a ring just greater than the diameter of the next-smaller mast section. The rings are used at the very top of a section, at G. Another design makes the hole in the ring just larger than the topmost swage outside diameter. This slides over the section, and rests on the shoulder of the swage, at G'. Another design makes the ring hole larger than the tubing. This ring is kept from sliding down the mast by a weld bead around the mast at A'. The ring located at G is the least desirable since the ring tends to jam if the mast is telescoped downward. The ring size varies with the section diameter.

There are two types of guy rings. One is a simple, flat plate, typically 1/8 inch thick. The other is thinner, with a lipped inner hole, and with the outer edge roll-crimped. It's claimed that the rolled edge eliminates the need for a guy thimble to keep the guy from chafing through. Prudence indicates the use of thimbles, which are *not* supplied with any masts I've seen.

These guy rings have holes for six guy wires, spaced as shown in Figure 2. They allow use of three sets of guys (commonly); four sets are rarely used.

Just above the bottom of the mast section, at H in Figure 1, is a weld ring. The size of this weld ring is controlled so that its outside diameter is just smaller than the inside diameter of the next-larger section. This acts with the swage of A or A' to take the bending moment when extended. It also prevents the inner tube from extending past the swage at F, preventing overextension.

At J is a pair of notches in the tube bottom. These are half circles, 5/16 or 3/8 inch across. They're intended to receive the cotter pin and prevent the mast section from rotating. The notches also distribute the shear load presented by the weight of the upper parts of the mast system.

After the sections are finished—with all holes punched or drilled and all welds made—they are hot-dipped galvanized. (Or at least, they should be!) I've seen one mast of unknown manufacture that was zinc-plated rather than hot-dip galvanized. I've been told paint-dipped masts exist. Such treatments are *not recommended,* as rust makes for a short mast life.

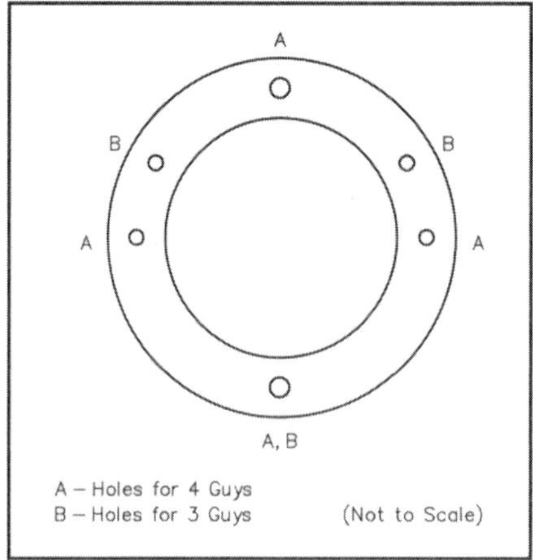


Figure 2—Top view of the guy ring plate found at the top of each section of a telescoping mast. The hole arrangement allows use of three or four sets of guys.

### Mast Specifications

Telescoping masts are manufactured in three weights. The light-duty mast is formed from 18-gauge steel, which has a nominal

**Glossary of Terms Used in the Tables**

- ALLOW—allowable load.
- AREA—Section area exposed to wind.
- A WND LD—allowable wind load.
- BOLT LD—maximum shear load on the bolt.
- COTR LD—maximum shear load on the cotter pin.
- DIAM—outside diameter of a section.
- GUY LEN—guy length.
- GUYS—length of three guys.
- GUY COMP—compressive load on a mast due to guy tension.
- GUY LD—guy load.
- GUY STR—guy strength.
- H WND LD—horizontal wind load.
- HGT—overall height.
- LAT AREA—cross section of tubing.
- LEN—length.
- LEN 3 GUYS—length of three guys.
- LEN 4 GUYS—length of four guys.
- LIMT LD—section load at failure limit.
- MAX AREA—maximum area of antenna and rotator.
- MAST WND—mast wind load.
- MAX V LD—maximum vertical load on section.
- NOM WT—nominal weight of a section.
- PANEL—a section of mast.
- RAD GYR—radius of gyration of section.
- SAFE LD—maximum safe load (with a safety factor of 4).
- STAT LD—static load.
- THICK—thickness of the section wall.
- TRY—a trial value.
- V WND LD—vertical wind load.
- WIND—wind load on projected area.
- WND LD—dynamic wind load.
- WT—weight.
- WT MAST—weight of the mast.
- WT GUYS—weight of the guys.

**Table 1**
**Telescoping Masts**

| *Quantity* | *Unit of Measure* | *Section* 1 | 2 | 3 | 4 | 5 |
|---|---|---|---|---|---|---|
| LEN | ft | 10 | 10 | 10 | 10 | 10 |
| HGT | ft | 10 | 19 | 28 | 37 | 46 |
| PANEL | ft | 9 | 8 | 8 | 8 | 9 |
| THICK | in. | 0.05 | 0.05 | 0.05 | 0.05 | 0.05 |
| DIAM. | in. | 2.25 | 2.00 | 1.75 | 1.50 | 1.25 |
| AREA | in.$^2$ | 0.35 | 0.31 | 0.27 | 0.23 | 0.19 |
| GUY LEN | ft | 25.08 | 29.83 | 36.24 | 43.57 | 51.43 |
| LEN 3 GUYS | | | | | | 558.43 |
| LEN 4 GUYS | | | | | | 744.57 |
| LAT AREA | ft$^2$ | 1.88 | 1.67 | 1.46 | 1.25 | 1.04 |
| NOM WT | lb | 15 | 13 | 12 | 10 | 8 |
| WIND | lb | 37.50 | 33.30 | 29.20 | 25.00 | 20.80 |
| RAD GYR | in. | 0.98 | 0.87 | 0.76 | 0.66 | 0.55 |
| MAX V LD | lb | 4070.26 | 3659.60 | 2903.26 | 2176.73 | 1281.30 |
| *With Cotter Pins* | | | | | | |
| COTR LD | lb | 745.14 | 745.14 | 745.14 | 745.14 | 745.14 |
| LIMT LD | lb | 745.14 | 732.14 | 720.14 | 710.14 | 702.14 |
| SAFE LD | lb | 186.28 | 183.03 | 180.03 | 177.53 | 175.53 |
| *With Stainless-Steel Bolts* | | | | | | |
| BOLT LD | lb | 2943.75 | 2943.75 | 2903.26 | 2176.73 | 1281.30 |
| LIMT LD | lb | 2943.75 | 2930.75 | 2891.26 | 2166.73 | 1273.30 |
| SAFE LD | lb | 735.94 | 732.69 | 722.82 | 541.68 | 318.32 |

Characteristics of typical mast sections. At the top of the table are the dimensions and weights of the five sections of a typical 50-foot mast. Derived quantities of wind area, section radius of gyration and section strength considered as a column are shown. The bottom parts show the shear strength and safe vertical load on the section for two types of retaining pins.

thickness of 0.049 inch. The heavy-duty mast is 16 gauge, with a nominal thickness of 0.063 inch. There is also an intermediate-weight mast, which uses 16-gauge steel for the top section, and 18-gauge for all other sections. This reduces the weight and cost of the complete assembly; it does not reduce the carrying capacity of a complete "as supplied" mast appreciably, but it does reduce the safety factor.

Because it seems likely that most amateur installations need the best possible mast, only the tallest, heavy-duty type is considered here.

Section specifications from the catalog of one large manufacturer are shown in Table 1. (See the "Glossary of Terms" for an explanation of the abbreviations used.) This 5-section design is often called a "50-foot" mast, but has a maximum height of 46 feet. One foot of height is lost in the overlap between each two sections. The table gives the overall height, the section length exposed to wind, and the panel length, the distance where there is no added strength from overlap with higher or lower sections. The weight of each section and its radius of gyration is shown, as well as the projected area and the wind loading for a 20 lb/ft² wind. This corresponds to a wind speed of 70.7 mph. This is selected as adequate for short-term use, such as a Field Day installation. The wind loading should be increased for permanent installations. Acceptable design values are:

| | |
|---|---|
| SE Florida, Cape Hatteras | 50 lb/ft² |
| SE USA coasts, some other areas | 40 lb/ft² |
| Rest of the contiguous states | 30 lb/ft² |

Load capability data shown later is based on the 20 lb/ft² value.

### A Standard Installation

The telescoping mast is designed to have a set of guy wires reaching from the ground to the top of each mast section. There are, of course, many mast heights, arrangements of these guys, as well as many ways of mounting antennas on the mast, and many antenna sizes. To reduce the analysis to reasonable size, a standard antenna mast installation is assumed.

We'll use a five-section mast with guys extending from the top of each section to a common point, as shown in Figure 3. Assume the common point to be located one-half of the total mast height from the base of the mast. All antenna and rotator weight is assumed to be located at the top of the mast: There are no intermediate antennas.

The standard guys are seven-strand galvanized wire common in TV installations. These guys are specified to have a tensional breaking strength of 910 lb and weigh 31.8 lb per 1000 feet. However, guy strength was found to be one of the limiting factors, so other material was also tried, as described later. Guys were assumed to be nonelastic: not changing length under load. I also assumed a negligible pre-load on the guys. Very closely, this corresponds to the precept that the guys should look tight and feel loose.

The mast is assumed to be as described in Figure 3, and to be in new condition. Lacking other information, common material strength values are used.

### Basis for Analysis

The foregoing assumptions allow analysis of each mast section as if it is totally independent of the others. The basic force diagram for the top section is shown in Figure 4. The horizontal force to the left is the wind load on the antenna and rotator and on the mast section. At the right is the slanting guy tension under load. This load is resolved into horizontal and vertical components.

Additional vertical loads on the mast are the weight of the antenna and rotator, and the weight of the mast section. The wind load on the mast is resolved into two components, one-half at the upper-guy attachment point, the rest being at the next guy, and not affecting this section. The wind load on the guy is neglected. The horizontal component of guy tension is equal to the sum of the wind loads at the upper guy. The vertical component then appears as a compressive load on the mast, adding to the weight of antenna, rotator and mast section, and guy. The relation between these components are:

Compression/Wind load = Guy height ÷ Guy base. For the top section and the installation assumed, the compression is just twice the total wind load. Guy tension = SCORE (Compression × Compression +

**Table 2**
**Telescoping Mast 2**

| Quantity | Unit of Measure | *Weight, Antenna and Rotator, lb* 10 | 20 | 30 | 40 | 50 | 100 |
|---|---|---|---|---|---|---|---|
| WT MAST | lb | 9 | 9 | 9 | 9 | 9 | 9 |
| WT GUYS | lb | 0.5 | 0.5 | 0.5 | 0.5 | 0.5 | 0.5 |
| STAT LD | lb | 19.5 | 29.5 | 39.5 | 49.5 | 59.5 | 109.5 |
| MAST WND | lb | 5.2 | 5.2 | 5.2 | 5.2 | 5.2 | 5.2 |
| *1/8-inch Steel Guy and Cotter Pin* | | | | | | | |
| ALLOW | lb | 175.5 | 175.5 | 175.5 | 175.5 | 175.5 | 175.5 |
| GUY COMP | lb | 156.0 | 146.0 | 136.0 | 126.0 | 116.0 | 66.0 |
| TRY GUY | lb | 174.4 | 163.2 | 152.1 | 140.9 | 129.7 | 73.8 |
| GUY STR | lb | 227.5 | 227.5 | 227.5 | 227.5 | 227.5 | 227.5 |
| GUY LD | lb | 175.5 | 163.2 | 152.1 | 140.9 | 129.7 | 73.8 |
| V WND LD | lb | 157.0 | 146.0 | 136.0 | 126.0 | 116.0 | 66.0 |
| H WND LD | lb | 78.5 | 73.0 | 68.0 | 63.0 | 58.0 | 33.0 |
| A WND LD | lb | 73.3 | 67.8 | 62.8 | 57.8 | 52.8 | 27.8 |
| MAX AREA | ft² | 3.7 | 3.4 | 3.1 | 2.9 | 2.6 | 1.4 |
| *3/16-inch Dacron Line and Bolt* | | | | | | | |
| ALLOW | lb | 318.3 | 318.3 | 318.3 | 318.3 | 318.3 | 318.3 |
| GUY COMP | lb | 298.8 | 288.8 | 278.8 | 268.8 | 258.8 | 208.8 |
| TRY GUY | lb | 334.1 | 322.9 | 311.7 | 300.5 | 289.3 | 233.4 |
| GUY STR | lb | 687.5 | 687.5 | 687.5 | 687.5 | 687.5 | 687.5 |
| GUY LD | lb | 334.1 | 322.9 | 311.7 | 300.5 | 289.3 | 233.4 |
| V WND LD | lb | 298.8 | 288.8 | 278.8 | 268.8 | 258.8 | 208.8 |
| H WND LD | lb | 149.4 | 144.4 | 139.4 | 134.4 | 129.4 | 104.4 |
| A WND LD | lb | 144.2 | 139.2 | 134.2 | 129.2 | 124.2 | 99.2 |
| MAX AREA | ft² | 7.2 | 7.0 | 6.7 | 6.5 | 6.2 | 5.0 |

Load and wind area spreadsheet calculation of allowable antenna area for various antenna plus rotator weights. The top part of the table shows the dead weights and mast wind loads. The lower parts combine these vectorially to obtain the allowable wind force and the antenna area, for the two types of support pins used. Note the large difference in antenna weight in the last two columns.

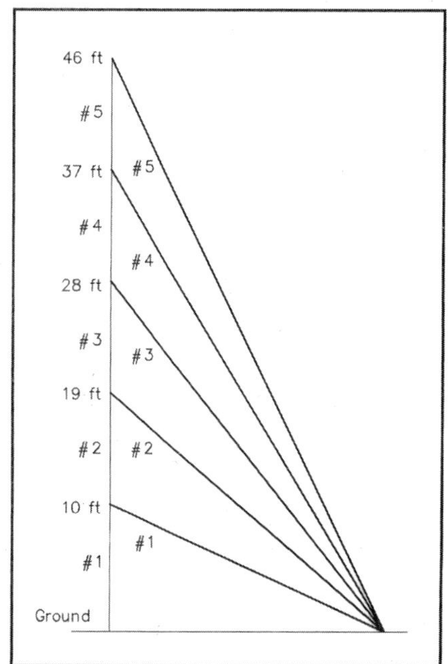


Figure 3—Assumed mast and guy installation. See text for assumption as to guy location.

tical antenna with no additions. Add a set of three to six wires parallel to the mast, and up to a foot or so from it, to form a cage for low loss and improved bandwidth. A bottom insulator of fiberglass cloth and resin can be used, or the cage wires can be fed as a folded dipole. A 3, 6, or even 10-foot capacity-hat is within the design capability of these masts. The usual rules about good grounds apply.

A simple installation for Field Day use replaces the top guys with a pair of 40- and 80-meter dipoles, with end cord added to keep the antenna ends as far above ground as the site permits. With a lightweight tribander at the top, scores will depend more on operators than on installation limits.

### Telescoping-Mast Installation

The telescoping mast is designed to be extended in only one way. *The mast must be vertical and the lower section properly guyed before extending the upper sections.* The bad reputation of the telescoping mast is almost completely due to attempts to ignore this simple rule. If you extend the sections, mount the antenna and then attempt to raise the assembly from a horizontal position to the vertical, you'll bend the mast. You may also damage the antenna and injure someone. In fact, the mast may bend if you try to raise it to the vertical while extended, just from its own weight—even if there's nothing mounted on the mast.

The recommended safe procedure is:

• Get a firm footing. For a typical Field Day installation on dirt, this means using a steel plate, with a spike on the bottom to penetrate the earth, and a pin on the top to keep the mast from slipping sideways. Base and tripod mounts for other surfaces are available.

• Mount the collapsed mast, plumb it to the vertical, and guy it with the bottom section permanent guys. A little pre-tension on these guys is a good idea.

• Tie one or two stepladders to the mast, for further work.

• Mount the rotator and antenna, and attach cables and the guys for the upper sections. Be sure that all fastenings are properly made and that the cables and guys won't tangle.

• If the antenna is small and light and there is no wind, an experienced person can get the antenna to a height of 30 feet. For greater heights, or with wind present, two to seven people are needed. Four handle the guys, two on ladders push up the sections and one person handles the clamping screws and the cotter pin or bolt. The guy handlers should not place strain on the guys, but should be prepared to keep the mast vertical. Practicing with the mast only, with no antenna in place, is a good idea.

• Starting with the top section, push it up a foot or two and tighten the clamp screw just enough to keep the section from slipping when released. Take another grip, and raise again, repeating until the stop is reached. Slip the cotter pin or bolt in place. Carefully lower the section until it rests on the pin or bolt. Turn the section until the slots engage. Then, tighten the clamp screw sufficiently to prevent rattling (wear a pair of good leather gloves). Don't use pliers—the screw will dent and deform the inner tube enough so that it will be impossible to lower the mast later.

• Repeat for the next lower section, and so on. If the mast starts to leave the vertical, or the wind picks up, stop, tie the guys temporarily and get help.

• Don't strain, and don't take chances. Even a few pounds falling from 10 or 20 feet can be deadly.

In principle, taking down a mast is easy: Just reverse the erection steps. If the mast is badly rusted, or the sections have been deformed by bending or by excessive clamping, however, this can be a chore—even impossible. If a little work doesn't get the antenna down to ladder height, consider bringing in a crane or ladder truck.

### Some Precautions to Take

Always check local building codes when considering the installation of a large antenna or high tower or mast structure. Much trouble and expense can be avoided by staying within imposed limits.

If you're going to use these telescoping masts above their rated TV antenna size and load limits, I recommend running a load analysis with the weights and dimensions involved. When doing this, include such factors as the distance from the top of the mast to the antenna mounting point, and the weight and wind area of the rotator. I recommend you *measure* the size and thickness of the mast sections. Errors in filling orders have occurred.

Remember: Safety First!

*Bob Haviland was first licensed as W9CAK in 1931. He obtained his BSEE degree at Missouri School of Mines in 1939, and his Professional Engineering license in New York.*

*Bob was project engineer for the first radio transmission from beyond the ionosphere, in 1949, at White Sands, New Mexico, and for the first missile launching from Cape Canaveral, in 1951. He developed ablation (the use of material which goes from solid to gas state) for space-vehicle reentry protection, and initiated programs for recovery of equipment and data from space.*

*Bob's worked extensively on communication and broadcast satellite concepts, and played a founding role in commercial satellite communication. Between 1959 and 1972, Bob was a member of US delegations to the ITU and CCIR. He served as Chairman, subcommittee for 27 to 1215 MHz, FCC WARC Advisory Committee for Amateur Radio, 1976.*

*Bob is a Fellow of the Institute of Electrical and Electronic Engineers, the American Astronautical Society and the British Interplanetary Society. He holds eight US patents, and is author of many articles and 14 books, including* The Quad Antenna, *CQ Publishing, 1993.* QST

# Contest Tips, Tricks & Techniques

Gary Sutcliffe, W9XT

## Portable Antennas and Supports

Selecting the best antenna and coming up with a method to get it up in the air can be a tough job. It's even tougher if the system also has to be portable and easy to set up. Whether for Field Day, a DXpedition or just a family vacation, a lot of ham ingenuity has gone into such systems. This installment looks into portable antenna systems as part of this special *NCJ* issue on portable operation.

### Antennas

A popular antenna is a $\lambda/2$ flat top at your lowest operating band fed with ladder line through a tuner. It can be used on all bands from the selected band on up. This antenna was mentioned by K5AF, K8DD and others. K9KM recommends twisting the ladder line feed line on these antennas to improve the balance. According to Howie, about one twist every 5 feet is enough.

A similar antenna is the G5RV, which can be used on multiple bands. On his trips to DX locations, K8DD takes two: a 51 footer and a 102 foot long version, along with a small tuner.

W2GD passed along his favorite portable antennas based on his 45 years of experience at Field Day. John likes the $\lambda/2$ 80 meter flat top feed with twin lead and a tuner for its flexibility. John will use it in inverted-V configuration if only a single high support is available.

For low angles on 80 and 160, W2GD likes vertical dipoles. He suggests trying to get one section as high as possible. The bottom half only needs to be above head height. John's other favorite is a 3 wire Yagi at 40 feet. The wires are in an inverted-V configuration. It is driven with 50 $\Omega$ coax.

KI5DI also likes 40 meter wire beams; he sent along information for a 2-element version. It uses two masts separated 16 feet for supports, while the driven element is essentially an inverted-V. The second mast supports a 71.5 foot wire reflector. More details can be found at Scott's Web site (**www.ki5dr.com/FD2002.html**).

Two wire antennas are sufficient for AK9F at this part of the sunspot cycle. Howard likes a standard 40 meter dipole fed with coax and a 135 foot flat top center fed with twin lead and a tuner. The flat top is used on both 80 and 20 meters. The 40 meter antenna goes up about 35 feet, and the other is around 65 or 70 feet; they are at right angles to minimize coupling.

Eric, K9GY, was very happy with a Force 12 Sigma-40XK/r which he used in the 2006 CW WPX. He also brought along a Cushcraft A50-3S. These fit into a Datrek hard case

Photos—TOM RUHLMANN, W9IPR

**Leon Rediske, K9GCF, adjusts one of the outrigger stabilizers on his tower trailer.**

**Hinged tower base plates built by Mark Potash, KC9GST.**

golf bag carrying case which made transportation easier.

Wire antennas tend to be the main type used by Gary, K9AY, for portable operations, due to the "rustic" locations from which he operates. Dipoles are used on 80, 40 and 20 meters, and delta loops are used on 15 and 10 meters. He also likes verticals for 80 and 40 meters, and when possible, uses both to increase the areas he can cover. Gary also notes that having portable antennas around can come in handy when he wants to put up a temporary antenna at home for a particular contest.

For the last 20 or so years, I have become a 40 meter CW specialist for Field Day. My favorite antenna is a dipole at about 35 feet with a wire reflector at about 7 feet — it is essentially a Yagi pointing up. This cloud warmer gives good coverage from high take-off angles down to the optimum angle for the East Coast from Wisconsin. Some hams just lay a reflector wire on the ground to do this, but my antenna modeling shows this has little effect. The 35 and 7 feet heights are convenient for supports and for keeping the reflector above head height.

I like a second antenna with a lower take-off angle for working the West Coast. A phasing network made from lengths of 50 and 75 Ω coax that lets me feed the antennas individually or simultaneously. A sloper from the top of a ridge worked well at a place I used to operate FD from, but I have not had much luck with them from my current location, and have been experimenting with different antennas. Vertically polarized antennas have proved to be too noisy from this location.

An interesting antenna was suggested by Paul, K5AF. It is an inverted-L, 33 feet up and 66 feet out. He either throws out some radials or uses his car as the ground! Dick, K4XU, stresses the importance of a good ground plane with cars. He notes that his mobile antennas perform much better with the flat bed of his truck as a ground plane than mag-mounted on his Jetta.

NE1RD likes using a fiberglass mast to support a 33 foot vertical radiator for 40 meters. Scott adds some elevated radials and feeds with coax; he also notes that it works on 15 meters. He has also used Buddipoles in dipole and vertical configuration, including using them from hotel balconies. Scott suggests whatever antenna system you use, to keep it simple. You are better off operating rather than spending a lot of time putting together a complex antenna system.

### Supports

If they are available, trees make excellent supports for wire antennas. A number of readers responded that trees are their first choice for supporting portable antennas. The first step is getting a rope into the right part of the tree to pull up the antenna. KJ9C prefers a bow and weighted arrow. Mel uses 8 pound fishing line on the arrow. This line is used to pull 30 pound line which in turn is used to pull up the rope. Mel can get a line up 100 feet and claims to have shot through a tower at the 60 foot level to pull through an extra guy line.

Another option for shooting a line into the trees is the wrist rocket, a form of sling shot. AK9F ties a line to a potato and uses a potato cannon to get a line up into the trees.

K9GY and others like the DK9SQ masts from Kanga US. These are telescoping fiberglass masts 10 meters long. K8DD has three; he uses them for dipole and G5RV support. On one trip, Hank used two to support a 40 meter half-square.

MFJ makes a similar mast used by K9CC. George doesn't extend it to its full length of 33 feet as he feels the top few sections are too small. He reports the mast will bend quite a bit, but if it bends too far, it will break with a noise similar to a gun shot.

A third option for fiberglass masts is suggested by K9AY. Gary likes the 28 foot fiberglass pole by Jackite (**www.jackite.com**). He says it is made for banners and such, and is less flexible than the ones by DK9SQ and MFJ.

A fishing pole from the discount rack is the support for N2WN's portable QRP operations when trees or other supports are not available. Julius picked up a lightweight 25 foot fiberglass rod that collapses down to 3 feet. He ties one end of a wire on the end eyelet, points it up in the air and connects the other end of the wire to the KAT2 tuner he uses with his K2. Julius says he has used this with success all the way down to the 80 meter CW band.

KH6LC has a lightweight two section crank-up tower that he uses when natural supports (such as trees) are not available. He can haul it in his truck, and a couple of people can put it up quickly; it can support a small beam. Lloyd says the biggest problem is driving the guy stakes into the solid Hawaiian lava.

"There is nothing like a good tower trailer," says N5OT. Mark uses one to support a 2-element KLM shorty-40 and a Cushcraft A4 that were modified for quick assembly and breakdown. Another advantage of the tower trailer is that you can travel with the antenna elements already together. Mark notes that they can be stored inside the tower. K4XU echoes Mark's remarks about trailer mounted crank-up towers. Dick says the ease of getting the tower up is even more important, with the average age in his Field Day group being more than 60.

The Ozaukee Radio Club (ORC), the local club I do Field Day with, has members with a couple of tower trailers, and they are great. One of them is pictured here. It is owned by Leon Rediske, K9GCF. The tower is hinged to fold down for travel, and a winch then cranks the tower up. Outrigger stabilizers slide out from the trailer, and cranks adjust the stabilizer feet to account for uneven ground.

The ORC has some nifty hinge plates for the base for the rest of their towers. These were made by Mark Potash, KC9GST, who owns a metal fabrication company. The plates are secured by driving rods into the ground. This is a great improvement over having the biggest ham in the group act as an anchor, keeping the tower in place while others walk it up. Mark also made some smaller versions for ground mounted verticals that have bolts for connecting radials.

For supports for my Field Day dipoles, I like the military masts made of tapered tubing. They come in sections around 4 feet long; you simply slide the tapered end into the bottom of the next section. Three people can easily put these masts up. I have even put them up by myself, but that is somewhat tricky.

I have three different styles. My favorite is about 2 inches in diameter and is made of heavy walled aluminum. It is good to 40 feet. I also have a similar style that is made of fiberglass. It is lighter and nice when you don't want a metal mast, but can't handle much side stress or the sides will split. The third style uses thin walled aluminum about 3 inches in diameter. It is light, but only good to about 25 feet.

There are a few dealers of the military mast at Dayton in the flea market each year. If you are lucky, you can find a complete kit including case, guy ropes and stakes. Otherwise, you just get the tubes. For attaching the guy ropes, you need something like a washer that just slips over the tapered part of the tube, but won't slide down further. I found the perfect parts at a farm supply store — I think they are some sort of flange for tractor mufflers. They even have the holes already punched to tie the guy ropes to!

That wraps up this installment of CTT&T. Hopefully you now have some new things to try for Field Day or other portable operations this summer. Thanks to AK9F, KH6LC, KJ9C, K4XU, K5AF, K8DD, K9AY, K9CC, K9GY, K9KM, NE1RD, N2WN, N5OT and W2GD for their comments. Thanks also to Tom, W9IPR, for supplying the photographs.

Next topic: Logging program display configuration. (Deadline May 15)

Most modern logging programs allow you to customize what information you see and where you put the windows with the information. What info windows do you show? How do you place them? Which ones do you find most useful? What new information would you like to see that is not supported — such as real time scoring on the competition or propagation info? Is there a danger of information overload? What contests do you use paper logs for? Why?

Send in your ideas on these subjects or suggestions for future topics. You can use the following routes: snail-mail: 3310 Bonnie Ln, Slinger, WI 53086. E-mail: **w9xt@unifiedmicro.com**. Be sure to get them to me by the deadline.

# Appendix

## Glossary of Terms

This glossary provides a handy list of terms that are used frequently in Amateur Radio conversation and literature about antennas. With each item is a brief definition of the term. Most terms given here are discussed more thoroughly in the text of this book, and may be located by using the index.

**Actual ground** — The point within the earth's surface where effective ground conductivity exists. The depth for this point varies with frequency and the condition of the soil.

**Antenna** — An electrical conductor or array of conductors that radiates signal energy (transmitting) or collects signal energy (receiving).

**Antenna tuner** — A device containing variable reactances (and perhaps a balun). It is connected between the transmitter and the feed point of an antenna system, and adjusted to "tune" or resonate the system to the operating frequency. It does not "tune" the antenna, only transform impedances.

**Aperture, effective** — An area enclosing an antenna, on which it is convenient to make calculations of field strength and antenna gain. Sometimes referred to as the "capture area."

**Apex** — The feed-point region of a V type of antenna.

**Apex angle** — The included angle between the wires of a V, an inverted-V dipole, and similar antennas, or the included angle between the two imaginary lines touching the element tips of a log periodic array.

**Azimuthal pattern** — The radiation pattern of an antenna in all horizontal directions around it.

**Balanced line** — A symmetrical two-conductor feed line for which each conductor has the same impedance to ground.

**Balun** — A device for feeding a balanced load with an unbalanced line, or vice versa. May be a form of choke, or a transformer that provides a specific impedance transformation (including 1:1). Often used in antenna systems to interface a coaxial transmission line to the feed point of a balanced antenna, such as a dipole.

**Base loading** — A lumped reactance that is inserted at the base (ground end) of a vertical antenna to resonate the antenna.

**Beamwidth** — Related to directive antennas. The width, in degrees, of the major lobe between the two directions at which the relative radiated power is equal to one half its value at the peak of the lobe (half power = –3 dB).

**Beta match** — A form of hairpin match. The two conductors straddle the boom of the antenna being matched, and the closed end of the matching-section conductors is strapped to the boom.

**Bridge** — A circuit with two or more ports that is used in measurements of impedance, resistance or standing waves in an antenna system. When the bridge is adjusted for a balanced condition, the unknown factor can be determined by reading its value on a calibrated scale or meter.

**Capacitance hat** — A conductor of large surface area that is connected at the high-impedance end of an antenna to effectively increase the electrical length. It is sometimes mounted directly above a loading coil to reduce the required inductance for establishing resonance. It usually takes the form of a series of wheel spokes or a solid circular disc. Sometimes referred to as a "top hat."

**Capture area** — See aperture.

**Center fed** — Transmission-line connection at the electrical center of an antenna radiator.

**Center loading** — A scheme for inserting inductive reactance (coil) at or near the center of an antenna element for the purpose of lowering its resonant frequency. Used with elements that are less than ¼ wavelength at the operating frequency.

**Choke balun** — A balun that works by presenting a high impedance to RF current.

**Coaxial cable or coax** — Any of the coaxial transmission lines that have the outer shield (solid or braided) on the same axis as the inner or center conductor. The insulating material can be air, helium or solid-dielectric compounds.

**Collinear array** — A linear array of radiating elements (usually dipoles) with their axes arranged in a straight line. Popular at VHF and above.

**Common-mode current** — Current that flows equally on all conductors of a group or that flows on the outside of the shield of a coaxial feed line.

**Conductor** — A metal body such as tubing, rod or wire that permits current to travel continuously along its length.

**Counterpoise** — A wire or group of wires mounted close to ground, but insulated from ground, to form a low-impedance, high-capacitance path to ground. Used at MF and HF to provide an RF ground for an antenna. Also see ground plane.

**Current loop** — A point of current maxima (antipode) on an antenna.

**Current node** — A point of current minima on an antenna.

**Decade** — A factor of ten or frequencies having a 10:1 harmonic relationship

**Decibel** — A logarithmic power ratio, abbreviated dB. May also represent a voltage or current ratio if the voltages or currents are measured across (or through) identical impedances. Suffixes to the abbreviation indicate references: dBi, isotropic radiator; dBic, isotropic radiator circular; dBm, milliwatt; dBW, watt.

**Delta loop** — A full-wave loop shaped like a triangle or delta.

**Delta match** — Center-feed technique used with radiators that are not split at the center. The feed line is fanned near the radiator center and connected to the radiator symmetrically. The fanned area is delta shaped.

**Dielectrics** — Various insulating materials used in antenna systems, such as found in insulators and transmission lines.

**Diffraction** — The bending of a wave by the abrupt edge or corner at a change in the medium through which the wave is traveling.

**Dipole** — An antenna, usually a half wavelength long, with opposing voltages on each half. Also called a "doublet."

**Direct ray** — Transmitted signal energy that arrives at the receiving antenna directly rather than being reflected by any object or medium.

**Directivity** — The property of an antenna that concentrates the radiated energy to form one or more major lobes.

**Director** — A conductor placed in front of a driven element to cause directivity. Frequently used singly or in multiples with Yagi or cubical-quad beam antennas.

**Doublet** — See dipole.

**Driven array** — An array of antenna elements which are all driven or excited by means of a transmission line, usually to achieve directivity.

**Driven element** — A radiator element of an antenna system to which the transmission line is connected.

**Dummy load** — Synonymous with dummy antenna. A nonradiating substitute for an antenna.

**E layer** — The ionospheric layer nearest earth from which radio signals can be reflected to a distant point, generally a maximum of 2000 km (1250 miles).

**E plane** — Related to a linearly polarized antenna, the plane containing the electric field vector of the antenna and its direction of maximum radiation. For terrestrial antenna systems, the direction of the E plane is also taken as the polarization of the antenna. The E plane is at right angles to the H plane.

**Efficiency** — The ratio of useful output power to input power, determined in antenna systems by losses in the system, including in nearby objects.

**EIRP** — Effective isotropic radiated power. The power radiated by an antenna in its favored direction, taking the gain of the antenna into account as referenced to isotropic.

**Elements** — The conductive parts of an antenna system that determine the antenna characteristics. For example, the reflector, driven element and directors of a Yagi antenna.

**Elevation pattern** — The radiation pattern of antenna at all vertical angles along a fixed direction.

**End effect** — A condition caused by capacitance at the ends of an antenna element. Insulators and related support wires contribute to this capacitance and lower the resonant frequency of the antenna. The effect increases with conductor diameter and must be considered when cutting an antenna element to length.

**End fed** — An end-fed antenna is one to which power is applied at one end, rather than at some point between the ends.

**F layer** — The ionospheric layer that lies above the E layer. Radio waves can be refracted from it to provide communications distances of several thousand miles by means of single- or double-hop skip.

**Feed line, feeders** — See Transmission Line.

**Field strength** — The intensity of a radio wave as measured at a point some distance from the antenna. This measurement is usually made in microvolts per meter.

**Front to back** — The ratio of the radiated power off the front and back of a directive antenna. For example, a dipole would have a ratio of 1, which is equivalent to 0 dB.

**Front to rear** — Worst-case rearward lobe in the 180°-wide sector behind an antenna's main lobe, in dB.

**Front to side** — The ratio of radiated power between the major lobe and that 90° off the front of a directive antenna.

**Gain** — The increase in effective radiated power in the desired direction of the major lobe.

**Gamma match** — A matching system used with driven antenna elements to effect a match between the transmission line and the feed point of the antenna. It consists of a series capacitor and an arm that is mounted close to the driven element and in parallel with it near the feed point.

**Ground plane** — A system of conductors placed beneath an elevated antenna to serve as an earth ground. Also see counterpoise.

**Ground screen** — A wire mesh counterpoise.

**Ground wave** — Radio waves that travel along the earth's surface.

**H plane** — Related to a linearly polarized antenna. The plane containing the magnetic field vector of an antenna and its direction of maximum radiation. The H plane is at right angles to the E plane.

**HAAT** — Height above average terrain. A term used mainly in connection with repeater antennas in determining coverage area.

**Hairpin match** — A U-shaped conductor that is connected to the two inner ends of a split dipole for the purpose of creating an impedance match to a balanced feeder.

**Hardline** — A type of low-loss coaxial feed line with a rigid or semi-rigid outer shield.

**Harmonic antenna** — An antenna that will operate on its fundamental frequency and the harmonics of the fundamental frequency for which it is designed. An end-fed half-wave antenna is one example.

**Helical** — A helically wound antenna, one that consists of a spiral conductor. If it has a very large winding length to diameter ratio it provides broadside radiation. If the length-to-diameter ratio is small, it will operate in the axial mode and radiate off the end opposite the feed point. The polarization will be circular for the axial mode, with left or right circularity, depending on whether the helix is wound clockwise or counterclockwise.

**Image antenna** — The imaginary counterpart of an actual antenna. It is assumed for mathematical purposes to be located below the earth's surface beneath the antenna, and is considered symmetrical with the antenna above ground.

**Impedance** — The ohmic value of an antenna feed point, matching section or transmission line. An impedance may contain a reactance as well as a resistance component.

**Impedance Matching Unit** — see Antenna Tuner.

**Inverted-V (dipole)** — A half-wavelength dipole erected in the form of an upside-down V, with the feed point at the apex. Its radiation pattern is similar to that of a horizontal dipole.

**Isotropic** — An imaginary or hypothetical point-source antenna that radiates equal power in all directions. It is used as a reference for the directive characteristics of actual antennas.

**Ladder line** — see Open-wire Line.

**Lambda** — Greek symbol (λ) used to represent a wavelength with reference to electrical dimensions in antenna work.

**Line loss** — The power lost in a transmission line, usually expressed in decibels.

**Line of sight** — Transmission path of a wave that travels directly from the transmitting antenna to the receiving antenna.

**Litz wire** — Stranded wire with individual strands insulated; small wire provides a large surface area for current flow, so losses are reduced for the wire size.

**Load** — The electrical entity to which power is delivered. The antenna system is a load for the transmitter.

**Loading** — The process of a transferring power from its source to a load. The effect a load has on a power source.

**Lobe** — A defined field of energy that radiates from a directive antenna.

**Log-periodic antenna** — A broadband directive antenna that has a structural format causing its impedance and radiation characteristics to repeat periodically as the logarithm of frequency.

**Long wire** — A wire antenna that is one wavelength or greater in electrical length. When two or more wavelengths long it provides gain and a multi-lobe radiation pattern. When terminated at one end it becomes essentially unidirectional off that end.

**Marconi antenna** — A shunt-fed monopole operated against ground or a radial system. In modern jargon, the term refers loosely to any type of vertical antenna.

**Matchbox** — see Antenna Tuner

**Matching** — The process of effecting an impedance match between two electrical circuits of unlike impedance. One example is matching a transmission line to the feed point of an antenna. Maximum power transfer to the load (antenna system) will occur when a matched condition exists.

**Monopole** — Literally, one pole, an antenna that operates with a single voltage with respect to ground, such as a vertical radiator operated against the Earth or a counterpoise.

**Null** — A condition during which an electrical unit is at a minimum. A null in an antenna radiation pattern is a point in the 360-degree pattern where a minima in field intensity is observed. An impedance bridge is said to be "pulled" when it has been brought into balance, with a null in the current flowing through the bridge arm.

**Octave** — A factor of two or frequencies having a 2:1 harmonic relationship.

**Open-wire line** — Consists of parallel, symmetrical wires with insulating spacers at regular intervals to maintain the line spacing. The dielectric is principally air, making it a low-loss type of line.

**Parabolic reflector** — An antenna reflector that is a portion of a parabolic revolution or curve. Used mainly at UHF and higher to obtain high gain and a relatively narrow beamwidth when excited by one of a variety of driven elements placed in the plane of and perpendicular to the axis of the parabola.

**Parallel-conductor line** — see Open-wire Line.

**Parasitic array** — A directive antenna that has a driven element and at least one independent director or reflector, or a combination of both. The directors and reflectors are not connected to the feed line. Except for VHF and UHF arrays with long booms (electrically), more than one reflector is seldom used. A Yagi antenna is one example of a parasitic array.

**Patch antenna** — A type of microwave antenna made from flat pieces of conductive material suspended above a ground-plane.

**Phase** — The relative time relationship of two signals.

**Phasing lines** — Sections of transmission line that are used to ensure the correct phase relationship between the elements of a driven array, or between bays of an array of antennas. Also used to effect impedance transformations while maintaining the desired phase.

**Polarity** — The assigned convention of positive and negative for a signal or system.

**Polarization** — The sense of the wave radiated by an antenna. This can be horizontal, vertical, elliptical or circular (left or right hand circularity), depending on the design and application. (See H plane.)

**Q section** — Term used in reference to transmission-line

matching transformers and phasing lines.

**Quad** — A parasitic array using rectangular or diamond shaped full-wave wire loop elements. Often called the "cubical quad." Another version uses delta-shaped elements, and is called a delta loop beam.

**Radiation pattern** — The radiation characteristics of an antenna as a function of space coordinates. Normally, the pattern is measured in the far-field region and is represented graphically.

**Radiation resistance** — The ratio of the power radiated by an antenna to the square of the RMS antenna current, referred to a specific point and assuming no losses. The effective resistance at the antenna feed point.

**Radiator** — A discrete conductor that radiates RF energy in an antenna system.

**Random wire** — A random length of wire used as an antenna and fed at one end by means of an antenna tuner. Seldom operates as a resonant antenna unless the length happens to be correct.

**Reflected ray** — A radio wave that is reflected from the earth, ionosphere or a man-made medium, such as a passive reflector.

**Reflector** — A parasitic antenna element or a metal assembly that is located behind the driven element to enhance forward directivity. Hillsides and large man-made structures such as buildings and towers may act as reflectors.

**Refraction** — Process by which a radio wave is bent and returned to earth from an ionospheric layer or other medium after striking the medium.

**Resonator** — In antenna terminology, a loading assembly consisting of a coil and a short radiator section. Used to lower the resonant frequency of an antenna, usually a vertical or a mobile whip.

**Rhombic** — A rhomboid or diamond-shaped antenna consisting of sides (legs) that are each one or more wavelengths long. The antenna is usually erected parallel to the ground. A rhombic antenna is bidirectional unless terminated by a resistance, which makes it unidirectional. The greater the electrical leg length, the greater the gain, assuming the tilt angle is optimized.

**Shunt feed** — A method of feeding an antenna driven element with a parallel conductor mounted adjacent to a low- impedance point on the radiator. Frequently used with grounded quarter-wave vertical antennas to provide an impedance match to the feeder. Series feed is used when the base of the vertical is insulated from ground.

**Sleeve balun** — A type of choke balun consisting of a ¼-wavelength metal tube or sleeve around a coaxial feed line that acts as an open-circuit to RF current.

**Stacking** — The process of placing similar directive antennas atop or beside one another, forming a "stacked array." Stacking provides more gain or directivity than a single antenna.

**Stub** — A section of transmission line used to tune an antenna element to resonance or to aid in obtaining an impedance match.

**SWR** — Standing-wave ratio on a transmission line in an antenna system. More correctly, VSWR, or voltage standing-wave ratio. The ratio of the forward to reflected voltage on the line, and not a power ratio. A VSWR of 1:1 occurs when all parts of the antenna system are matched correctly to one another.

**T match** — Method for matching a transmission-line to an unbroken driven element. Attached at the electrical center of the driven element in a T-shaped manner. In effect it is a double gamma match.

**Tilt angle** — Half the angle included between the wires at the sides of a rhombic antenna.

**Top hat** — See Capacitance hat.

**Top loading** — Addition of a reactance (usually a capacitance hat) at the end of an antenna element opposite the feed point to increase the electrical length of the radiator.

**Transmatch** — See Antenna tuner.

**Transmission line** — A cable that transfers electrical energy between sources and loads.

**Trap** — Parallel L-C network inserted in an antenna element to provide multiband operation with a single conductor.

**Tuner** — see Antenna Tuner

**Twinlead** — A type of open-wire line encased in plastic insulation for its entire length. See also Window line.

**Uda** — Co-inventor of the Yagi antenna.

**Unipole** — See monopole.

**Unun** — Unbalanced-to-unbalanced impedance transformer.

**Velocity factor** — The ratio of the velocity of radio wave propagation in a dielectric medium to that in free space. When cutting a transmission line to a specific electrical length, the velocity factor of the particular line must be taken into account.

**Vivaldi antenna** — A type of microwave antenna that uses an exponential cutout as the radiating element, similar to an exponential horn.

**VSWR** — Voltage standing-wave ratio. See SWR.

**Wave** — A disturbance or variation that is a function of time or space, or both, transferring energy progressively from point to point. A radio wave, for example.

**Wave angle** — The angle above the horizon of a radio wave as it is launched from or received by an antenna. Also called elevation angle.

**Wave front** — A surface that is a locus of all the points having the same phase at a given instant in time.

**Window line** — A type of twinlead with regular rectangular holes or "windows" in the insulation between conductors.

**Yagi** — A directive, gain type of antenna that utilizes a number of parasitic directors and a reflector. Named after one of the two Japanese inventors (Yagi and Uda).

**Zepp antenna** — A half-wave wire antenna that operates on its fundamental and harmonics. It is fed at one end by means of open-wire feeders. The name evolved from its popularity as an antenna on Zeppelins. In modern jargon the term refers loosely to any horizontal antenna.